Saulo Gonçalves Pereira

Environmental Education

Saulo Gonçalves Pereira

# Environmental Education

## Faces and Possibilities of a Possible Practice

ScienciaScripts

**Imprint**
Any brand names and product names mentioned in this book are subject to trademark, brand or patent protection and are trademarks or registered trademarks of their respective holders. The use of brand names, product names, common names, trade names, product descriptions etc. even without a particular marking in this work is in no way to be construed to mean that such names may be regarded as unrestricted in respect of trademark and brand protection legislation and could thus be used by anyone.

Cover image: www.ingimage.com

This book is a translation from the original published under ISBN 978-3-330-99753-0.

Publisher:
Sciencia Scripts
is a trademark of
Dodo Books Indian Ocean Ltd. and OmniScriptum S.R.L publishing group

120 High Road, East Finchley, London, N2 9ED, United Kingdom
Str. Armeneasca 28/1, office 1, Chisinau MD-2012, Republic of Moldova, Europe
Printed at: see last page
**ISBN: 978-620-8-26639-4**

**ENVIRONMENTAL EDUCATION: Faces and Possibilities of a Possible Practice**

**AUTHOR**
**SAULO GONQALVES PEREIRA**

**Abstract:** Environmental education is a set of educational processes, school-based or not, in which the individual and the community seek skills, social values, and competencies in the pursuit of environmental conservation and the quality of life of all individuals and their surroundings. This manuscript proposes to present a contextualization of environmental education, showing the faces and possibilities of a possible practice in the school (formal) and out-of-school (informal) contexts, as well as its characteristics and origins. It will deal with the initial concepts and introduction to environmental education, the approach to environmental preservation in the school context, sustainability and recycling in question, the role of the consumer and the future, environmental education and the treatment of effluents, environmental education as a subsidy for the conservation of the cerrado, environmental education and the preservation of the cerrado. The discussion does not end with the presentation proposed here, because the search for knowledge about environmental education is broad and cyclical. This work is recommended for the community in general, as well as students, teachers of basic and higher education, and environmental researchers.

## Preface

It is an immeasurable pleasure to preface this manuscript, which focuses on environmental education. I believe that talking about environmental education is expressing myself about something I believe in very strongly, and that is life itself and its infinite relationships between different beings and between them and the environment. It is extremely necessary to reflect on this issue, seeking out its causes and effects, pointing out solutions, encouraging participation, as well as strengthening groups and individuals at regional and local levels in the search for a better world for everyone, so that each person feels able, motivated and responsible for making it a better world for all beings. Today, we are living through a turbulent process of urbanization and growth that originated during the industrial revolution at world level. In Brazil, this process began, above all, at the end of the 1960s. These changes have made the Earth a difficult place to live for some organisms as a result of the unregulated exploitation of others! It is important that we all think about this process together and come up with solutions for the common future. The problems arising from this model are numerous and the effects are negative, so environmental education comes precisely to show that human beings are capable of generating significant changes by walking along paths that lead to a more socially just and ecologically sustainable world. This manuscript was organized based on the experiences of a team of students who wrote their final course papers on environmental education in their own realities.

The manuscript discusses the introduction to environmental education, the approach to environmental preservation in the school context, sustainability and recycling, the role of the consumer and the future, environmental education and the treatment of effluents, environmental education as a means of conserving the cerrado. This manuscript was produced by compiling various works which I had the opportunity to guide and participate in as a reviewer.

I would like to congratulate all environmental agents and all researchers into environmental education, especially the co-authors of this manuscript for their clear and pertinent exposition of environmental education in the world, in Brazil and in Minas Gerais. Finally, I would like to point out that it is the reader who will be able to make the most of this manuscript.

**Thanks to co-authors, collaborators and advisors**

DANIELA CRISTINA SILVA BORGES

WANDERSON ALVES PEREIRA

DENIZIA CRISTINA SILVA

LEANDRO JUNIO DA COSTA FERREIRA

MARTA NUNES

ROSELINE MARTINS SABIAO

HELEN PAULA RODRIGUES DE QUEIROZ

MARCIO EDIRLEI DA SILVA

CAMILA MARTINS SILVA

GISLEINE DE SOUZA OLIVEIRA

RAISLA BIANCA SILVEIRA

**SUMMARY**

# 1. ENVIRONMENTAL EDUCATION AND SUSTAINABILITY

## 1.1 Environmental education and its concepts

Educating is an act of extreme necessity, an ontological virtue of our nature, which must be understood if it is to be carried out successfully (LOUREIRO, 2006).

Environmental education was legally instituted in Brazil by federal law No. 6938, passed on August 31, 1981, when the National Environmental Policy (PNMA) was established. This law is a historic fact in the institutionalization of the defence of Brazil's environmental quality. In addition, the National Environmental System (SISNAMA) was created in order to ensure that all the levels of action, especially governmental ones, were organized in the same law.

The priority of environmental education should be critical and innovative thinking, in any time, place or circumstance, in its formal, non-formal and informal forms, generating the modification and construction of society (AVANZI, 2004).

Recent studies show that the environment is a topic of concern for people when they remember to see it in its broadest sense and as an element that enables human survival, in an interaction that is not exhausted, since the environment establishes the interrelationships and interdependence of the various elements in the constitution and maintenance of life.

There are many challenges to be faced when trying to direct action towards improving living conditions in the world. One of them is interaction with the basic asset for human life: the environment. Environmental education can take place in schools, companies, universities, public offices, among others. This education can be extended by government bodies or by environmental organizations (IBEMA, 2003).

The practices grouped under the definition of environmental education have been categorized in many ways: popular, critical, political, community, formal, non-formal, environmental education for sustainable growth, conservationist, socio-environmental, outdoor, for the solution of many problems (CARVALHO, 2004).

It is difficult, according to Jacobi (2003) [...] "to solve the growing and complex environmental problems and reverse their causes without a radical change in the systems of knowledge, values and behaviors that are generated by the existing dynamics of rationality".

As the above-mentioned authors have said, the main objectives of environmental education are to make everyone aware of the need to care for the environment and to develop growth that, through new generations, has the least possible impact on the environment, but this practice is still not widespread.

Education contributes directly to changing the current patterns of use and distribution of environmental goods towards more sustainable, fair and solidary ways of living and relating to nature (CARVALHO, 2004). In this way, the theme is associated with a number of contexts, such as sustainability and recycling. Sustainability suggests an indispensable interrelationship between social justice, environmental balance, quality of life and a break with the current pattern of development (JACOBI, 2003). Recycling, on the other hand, addresses the contexts of reuse, recycling and reflections on consumerism.

Environmental education should be dedicated to society as a whole, given the worsening environmental crisis for sustaining life on the planet and the need to tackle it (GUIMARAES, 2005). Environmental education must help to raise ethical awareness of all the life forms with which we all share planet Earth, respect their vital cycles and impose limits on the exploitation of these life forms by human beings (FORUM GLOBAL, 1992 apud GADOTTI, 2000).

An educational process to change the vision of the history of exploitation of natural resources will deal with awareness that includes a totality in action. The strength and significance of everyday actions underpin or subsidize major environmental undertakings, such as reordering consumption, changing relationships, selective collection and recycling, combating poverty, basic sanitation, among others, in order to tackle recurring environmental problems (RUSCHEINSKY, 2004a). In this way, through education everyone can become aware that by being less consumerist, having orderly growth with social equality for all people, it is possible to reduce today's environmental problems.

Environmental education has received considerable impetus, gaining increasing public recognition through reflections and actions promoted by a variety of agents from international organizations, governmental and non-governmental organizations, social movements, universities and schools (LIMA,

2006).

Environmental education, understood from the perspective adopted, must be carried out by bringing together formal and non-formal educational spaces; by the complicity of the school and the community; by the collective and popular construction of the political-pedagogical project and by linking cognitive activities with changes in life practices (LOUREIRO, 2006).

There is a need to point out the concepts that guide environmental education proposals through a critical distinction of their political and ethical meanings, so that they can help educators and students in their choices and positions on reality (LIMA, 2008).

Environmental education can be briefly divided into three parts: critical, transformative and emancipatory. Critical insofar as it debates and points out the contradictions of the current model of civilization, the inclusion of society in nature and the social relations it establishes. Transformative, because by debating the nature of the civilizational process, it stimulates humanity's capacity to build another future from the construction of another present and thus establish new relationships between human beings and nature. It is also emancipatory, because it takes freedom as a fundamental value and seeks to produce autonomy for subaltern, excluded groups.

In Brazil, the environment at school has been the subject of complex pedagogical debates in recent times, leading to deep insecurity, especially among teachers and future teachers working in schools. A few years ago, for the vast majority of teachers of various subjects, the prevailing view was that the school's main role was to deal with the subject in a superficial way, far removed from the student's reality; the subject was only the focus of the disciplines of Science, Geography and related studies on Earth. However, this assumption has been destroyed as man has seen the changes in the climate, the responses that nature has given to human neglect of the natural elements. The current curricular parameters point to the need for interdisciplinarity in the study of the cross-cutting theme, but in a superficial way and with little indication and practice.

Therefore, research that seeks to verify the socio-environmental crisis shows that environmental education is an inevitable solution to tackling this problem. The educational process is perceived by society as a space for its realization, and this is the place where initiatives to include environmental education in the school agenda are moving, albeit timidly. Schools are not disconnected from society and are also experiencing this paradigm crisis. Knowledge of environmental education networks is a strategy with the capacity to break down the isolation of education actors, making them part of a group movement; providing interconnection of local reality from an expanded perspective; and experiencing an educational environment for continuous formation.

Over the course of history, man has changed as a result of environmental action, he has created culture, economic options, communication with the environment and with others. But it is necessary to reflect on how these socio-economic and environmental relationships should be, and thus seek appropriate decisions for quality of life and environmental balance (BRASIL, 2000).

Therefore, according to the National Curriculum Parameters (PCN), the main function of working with the environment is to contribute to the formation of aware citizens who are able to decide and act in socio-environmental reality in a way that is committed to life, to the well-being of each individual and of society as a whole. To achieve this, schools need to work with attitudes, training and values, teaching and learning skills and procedures, rather than just information and concepts. This is a major challenge for education. "Environmentally correct" behavior will be learned in day-to-day practice at school: gestures of solidarity, personal hygiene habits in different environments, participation in small negotiations can all be examples of this.

The development of citizenship and the formation of environmental awareness has in the school a suitable place for its realization through active and participatory teaching, capable of overcoming the impasses and dissatisfactions experienced in general by the school today, based on traditional ways.

We need to take a transformative step. This step points in the direction of orienting school work towards an environmental logic, in order to move from an informative school to a formative school. It is necessary and possible to contribute to the formation of people capable of creating and expanding spaces for participation in "decision-making" on environmental problems.

Therefore, Environmental Education seeks to encourage individuals to actively participate in solving problems in their specific context of reality, starting from theoretical and practical data in a systematic and holistic way based on integrated studies of the society/nature relationship, seeking regional sustainability, because according to Guattari (1990), only "ecosophy", an articulation between what he considered "the three ecological registers" - the physical environment, social relations and human subjectivity - can deal with environmental issues. The key role of teachers in environmental education also recognizes the need to introduce environmental education into teacher training courses, an initiative spearheaded by EEITE *(Environmental Education into Initial Teacher Education in Europe),* which has characteristics that differ according to the different situations and educational levels of each country involved. In Brazil, even after Rio-92, the reductionist disciplinary approach seems to persist, according to the group studied, despite the interdisciplinary approach advocated since 1977 in Tbilisi (UNESCO, 1997), a phenomenon that Giordan (1994) points out is universal, mainly due to rigid curricula.

Environmental issues have emerged in contemporary society, permeating various fields of knowledge and having a significant impact on the aims of school education. In this sense, Environmental Education is often pointed to as one of the main ways of training people capable of dealing with socio-environmental problems and conflicts, so that they are prepared to understand the scientific and technological debates involved in these issues.

Formal education, by maintaining overloaded teaching hours and a curriculum organized in a disciplinary manner, does not provide interdisciplinary experiences, as health education and environmental education require. This contributes to the fact that local problems, even if they are known by many teachers, are not seen as issues to be tackled in the classroom. This is perhaps because such problems involve too many variables in the study of their complexity, or because teachers lack the motivation to deal with problems that are felt to be beyond their strength, especially if they work in isolation (GRYNSZPAN, 1999, p. 5).

The recent legislation on environmental education (BRASIL, 1999) points to a national policy that values the training of human resources *aimed at "developing instruments and methodologies, with a view to incorporating them, in an interdisciplinary way, into the different levels and modalities of teaching".* In addition, we propose the development of strategies to establish an effective culture of continuous teacher training. This would allow, for example, work with teachers centered on thematic axes. An integrative approach for health education and environmental education would be the "quality of life" axis, with a deeper understanding of the concept of balance and its history being essential for this pedagogical approach. This perspective is also in line with the 1997 National Curriculum Parameters (BRASIL, 1997).

Education can't provide material support, such as providing vaccines or building sewage systems, but that doesn't mean it has to be seen as secondary, complementary, but not the core, as is usually the case. As its primary mission is the transmission of values, it is fundamental for tackling life's issues through the formation of critical subjects. Etymologically, it means fostering the flowering of something that is already within the person and not filling an empty vessel with knowledge. For this reason, emphasis is placed on the representations and strategies brought about by this experience.

For Meyer (1998), the three educational processes proposed by Lucas (1980) *on, in* and *for* the environment, have gradually been replaced by each other because this is a process that has required reflection to reach the level of the current proposals of many schools and many ecological associations that accept and propose a *triple approach to* environmental education, or, as we might call it, a *complex approach to* environmental education.

On the other hand, Sanmarti (1994, p.45) states that "[...] the results of an educational process are not the consequence of a single activity, but of a prolonged educational action over years, i.e. there is no cause and effect relationship, but mutual influences between knowledge, attitudes and behaviors".

It is therefore understood that, according to the PCN's, environmental issues constitute a cross-cutting theme which, although it is most clearly present in the disciplines of Science, History and Geography, also needs support from the areas of Mathematics, Portuguese Language, Physical Education and the Arts in order to be understood. The teacher must show that, in order to become aware of environmental issues, students will have to engage in constant learning, as natural transformations also take place on an ongoing basis.

The teacher's main concern should be to develop values, attitudes and ethical attitudes, because the concepts needed to deal with environmental issues are learned in other subjects. Show your students the differences between balanced, healthy environments and polluted or degraded places. It's important for them to realize that seeing something wrong is not a reason to be discouraged, but to mobilize the school and the community to solve it (PEDRON; ROSA, 2011, p.315). It is necessary for the school to deal with issues that interfere with students' lives and with which they are confronted on a daily basis.

From this perspective, social issues relating to ethics, health, the environment, cultural plurality, sexual orientation and work and consumption are integrated into the educational proposal of the National Curriculum Parameters as Transversal Themes. They are not new areas, but a set of themes that appear transversalized, permeating the conception of the different areas, their objectives, contents and teaching guidelines.

Transversality presupposes an integrated treatment of the areas and a commitment to interpersonal relationships within the school, since the values that are to be transmitted, those experienced in school life and the coherence between them must be clear in order to develop the students' ability to intervene in reality and transform it.

This characteristic seems to fit in with one of the aims of HE, which is to build knowledge without reducing it to disciplinary schemes. In this respect, Shulman (1992) adds that cases have the potential to teach theoretical principles or concepts, precedents for practice, moral or ethical principles, strategies for certain professional practices, visions or images of the possible, as well as reducing the distance between reality and theoretical knowledge. In view of the above, it can be seen that currently, HE is one of the main emphases of curricular proposals and is associated with the formation of citizenship, considered one of the main trends increasingly encouraged during the 1990s (LAYRARGUES, 1999).

The theme "environment", included in the National Curriculum Parameters (PCN), aims to "contribute to the formation of conscious citizens, able to decide and act in the socio-environmental reality in a way that is committed to life, to the well-being of each individual and of local and global society" (BRASIL, 1998a, p. 11).

It should also be pointed out that it contributes to the development of critical thinking, based on contemporary concepts, favoring a participatory stance among these teachers in their identities and which could contribute to improving the quality of life, a common objective of environmental education. As Sorrentino (1991, p. 02) advocates, participation is [...] *"the purpose and viability of education,* since it helps to overcome the distancing attitude to which, in a way, we are relegated by a confluence of factors in so-called modern life".

In this way, we emphasize the belief that an educational project is more than just training and knowledge of the facts, but that it encourages the search for new sources and solutions. The persistence of traditional, abstract and compartmentalized basic education has not encouraged the analysis of local problems. Furthermore, environmental education and health education are still considered to be the responsibility of science teachers in reality, despite official recommendations.

On the contrary, we should follow the line that has been emphasized by Giordan (1994). We also insist on the shared construction of knowledge, or values, based on the rich exchanges derived from intra-group interactions. Creativity, imagination and daring must flourish in the "classroom" environment. Innovative initiatives to improve this globalized world through the problematic: the environment and contemporary reality.

## 2 HISTORY OF ENVIRONMENTAL EDUCATION

The term "Environmental Education" was only coined in the 1970s, mainly when the need arose to reflect on environmental issues (BRASIL, 2007). Since then, various episodes have followed, and these issues have been consolidated. Various conferences, debates, symposiums and national and international movements in favor of caring for the Earth have taken place over time.

"The 1960s can be considered a benchmark for the origin of concerns about the loss of environmental quality" (TOZONE- REIS, 2004, p. 3). However, it was from 1972 onwards that studies and proposals on environmental education were introduced in various countries, including Brazil. International conferences organized by the UN (United Nations Organization) were held, with representatives from various countries coming together to define and create proposals on the environment. The conferences were the starting point for spreading environmental concern and protection around the world. Of course, each country has its own way of dealing with environmental issues.

From June 5 to 16, 1972, in Sweden, representatives from 113 countries took part in the Stockholm Conference on the Human Environment, which is said to be the first conference/meeting on the environment held by the UN. During the conference, the "Declaration of the United Nations Conference on the Human Environment" was established, which stipulated axes to be fulfilled by nations in order to reduce pollution rates and preserve the environment. However, there was a conflict between the developed and underdeveloped countries, with representatives of the developing countries accusing the developed countries of wanting to limit their growth with the excuse of environmental preservation. "To the astonishment of the world, representatives from Brazil called for pollution, saying that the country would not mind paying the price for environmental degradation as long as the result was an increase in GNP (Gross National Product)" (DIAS, 2004, p. 36).

In 1977, in the former Soviet Union, the Tbilisi Conference took place, known as the First Intergovernmental Conference on Environmental Education, which discussed ways of incorporating the environmental dimension into all forms of education. The Tbilisi conference is a continuation of the one held in Stockholm in 1972, and is considered the starting point of the International

Environmental Education Program. The conference established objectives, principles and strategies for environmental education around the world.

In 1992, from June 3 to 14 in Rio de Janeiro, the United Nations Conference on Environment and Development, known as the Rio-92 Conference, was held with the participation of 170 countries. The Rio-92 conference was a milestone in history, because it admitted the unsustainability of the development model, the representatives of the countries admitted that something had to be done, that sustainable measures had to be adopted.

In 2002, the Rio+10 conference was held between August 26 and September 4 in Johannesburg, South Africa.

> The UN promoted the World Summit on Sustainable Development, which became known as Rio+10. Representatives from 189 countries took part in the event, as well as hundreds of non-governmental organizations. The purpose of the conference was to discuss the degree of anthropogenic (human) interference in the environment, and for the first time globalization was associated with the problems. As a result, a document was produced with implementation plans for environmental protection, combined with economic and social development (DINIZ, 2002).

The environmental future of the planet was in the hands of the leaders of all the world's nations. Never before have we heard so much about environmental issues as recently, thanks to the *Copenhagen* Summit (Denmark), which took place between December 7 and 18, 2009. Issues related to global warming and the targets for reducing the emission of polluting chemical gases were discussed.

The *Copenhagen* Summit *has come to rectify (again!), among others, the Rio de Janeiro Summit in 1992 (ECO 92) and the* Kyoto Summit *in* 1998. Rectifying agreements that should already be fully operational. But it has been clear for a long time that there is a deeper problem: no one, i.e. the countries, are keeping their promises to cut emissions.

It is believed that this issue is extremely important and urgently needs to be addressed, since the problems caused by the beginning of the warming process are evident in our daily lives. Since when have we heard of so much heat? Or such strong storms? Some say that these are seasonal storms. It would be acceptable to think so, but the world's population has almost doubled in 7 decades, which means that more people are affected, especially those living in extreme poverty.

Brazil is taking an irresolute stance: at first it talks about reducing the degradation of the Amazon, and then it talks about loans to international investment and financing funds. It's a contradictory idea, because if there were a good basis, especially educational, it would be possible, but it's not the reality. The problem, according to *"People forget that until recently there wasn't even a consensus within the government on reducing deforestation,"* (GUIMARAES; FONTOURA, 2012). This implies that these are goals that depend on awareness and time, and time is running out!

This game of "push-pull" and this position of being "on the fence" does nothing to help us make strong decisions. And our politicians, who are divergent, presented various opinions within the same delegation, and gaffes were heard that will not be easily forgotten. This demonstrates the fragility of a nation that retains the world's greatest biodiversity and around 20% of the world's water resources. What is most intriguing is the discussion about the *Pre-Salt (*which is a major non-renewable and polluting energy source) where Brazil, which is "so committed" to reducing pollution and renewing energy resources, is fixated on this (undeniable) economic opportunity and still manages to defend the idea of environmental policies.

In 2012, the United Nations Conference on Sustainable Development, better known as the Rio+20 conference, was held in Rio de Janeiro, Brazil, from June 13 to 22. Representatives from 193 countries took part in this event, which aimed to analyze sustainable development and the changes that have taken place since Rio-92 [...] "never has humanity been so close, not to the disaster that many apocalyptic environmentalists suggest, but to experiencing the consequences of the fragility of the systems that are vital for life on the planet. " (GUIMARAES; FONTOURA, 2012).

Twenty years after Eco92, in June 2012 Rio de Janeiro welcomed government leaders and civil society from various countries to discuss plans and actions for the future of the planet. The United Nations Conference on Sustainable Development should have reached conclusions for a common agenda on the environment for the coming decades, but it didn't work out that way, not by a long shot!

However, for many, the RIO+20 was not even known about, and many of the people I spoke to about it either didn't know what it was about or didn't take such an important issue seriously. For the more informed, the results were disappointing, especially for environmental organizations and international leaders. On the other hand, representatives of the Brazilian government say they considered it a success. The UN, with its text full of promises to eradicate poverty without harming nature and all the rest, has fine print, but few targets, which is what matters.

The aim of the Rio + 20 convention was to implement sustainable development throughout the world in order to improve the Earth's environmental conditions, so that future generations can have an ecologically balanced environment, a good for the common use of the people and essential to a healthy quality of life. The debates affirmed, for example, that farmers can make important contributions to sustainable development, as can industries.

It is believed that the activities, provided they are conducted in a sustainable manner, will lead to increased food security, thus ensuring economic growth and the livelihoods of the poorest people in all parts of the world. In the end, it was agreed that there was no agreement. The end result was that the responsibilities of each country and the deadlines for improvement were not identified, nor were sources of funding for actions aimed at sustainable development, especially in developing countries. The summit's final document, entitled "The Future We Want", is full of good intentions and few concrete actions.

The Paris Conference was held in 2015, with the long-term aim of preventing the planet's temperature from rising by more than 2°C, in order to avoid disastrous effects.

> This means taking action on multiple fronts. And it has been concluded that it is necessary to encourage the development of clean energies and reduce the use of fossil fuels. But beyond the tangible threats, global warming has already become an economic issue that divides and pits rich and poor countries against each other. Developing countries, which are more vulnerable to the effects of climate change, are unwilling to take part in the common struggle unless the richest (and most responsible for emissions) commit to offering financial incentives for the former to tackle global warming, i.e. without conclusions (INSTITUTO SOCIOAMBIENTAL, p.08).

On May 2, 2017, the United States of America, through its president Donald Trump, withdrew from the agreement, with the rhetoric "development", they need to "develop the economy", causing a lot of insecurity in the world.

## 3 FORMAL ENVIRONMENTAL EDUCATION

At the Rio/92 International Conference, representatives from more than 170 countries signed treaties recognizing the essential role of education in building an ecologically balanced world, which requires individual and collective responsibility at all social levels. In Brazil, Environmental Education was made a national obligation by the Constitution promulgated in 1988 (BRASIL, 2000).

Environmental education became law on April 27, 1999. Law No. 9.795 - Law on Environmental Education, instituting a national policy on environmental education. Brazil is the only country in Latin America to have a specific national policy on environmental education (DIAS, 2004).

According to the National Environmental Education Policy (Law No. 9.795/99), environmental education should not be implemented as a specific subject in the teaching curriculum. Article 2 states that:

> Environmental education is an essential and permanent component of national education, and must be present, in an articulated manner, at all levels and modalities of the educational process, both formal and non-formal" (DIAS, 2004, p. 202).

Art. 9 Environmental education in school education is understood to be that developed within the curricula of public and private educational institutions, encompassing, according to Dias, (2004, p. 02):

- basic education
- early childhood education;
- elementary school e;
- High school;
- higher education
- special education;
- professional education;
- youth and adult education.

> Art. 10 Environmental education will be developed as an integrated, continuous and permanent educational practice at all levels and modalities of formal education (DIAS, 2004).

Brazil's environmental education law is very complete and meets all the parameters, but it is up to teachers, the government and environmental organizations to better evaluate the ways in which this law can be introduced. After 18 years in which Law 9.795 was implemented, the objective has not been achieved. In practically all of Brazil, resources are being exploited without the slightest protection. Most Brazilian individuals do not fulfill their role in environmental preservation and the sustainable use of resources. What can be done to change reality? Why do we have such a good law on paper, but no considerable results? Unfortunately, this question is still unanswered.

Environmental Education must be included in all forms of teaching, through a participatory pedagogical process that seeks to instill in students a critical awareness of environmental problems [...] in this way, schools must increasingly include and develop environmental education in their teaching practices, whether through projects or program content (DIAS, 2004).

Within the CBC (Common Basic Content) for elementary school science from the 6th to the 9th grade, it can be analyzed in Thematic Axis I, theme 1 (Diversity of life in the Environments) that the proposal for Environmental Preservation is requested by the Department of Education very superficially. Most of the topics are not compulsory but suggestions for the teacher to work on. Theme 5 (Water Quality and Quality of Life) asks for a very in-depth approach to water, its importance and its preservation. Oliveira (2007, p. 480) states that:

> We found indications that there is a need for more theoretical and methodological support for science teachers to teach and promote human encounters with the environment. Students can only become researchers through well-planned contact with nature, so that they realize that it is not just another theme or chapter in the textbook. Students need to see themselves as part of the environment from an early age. Too often, fragmented content is seen without the proper connections being made, taking into account only the sequence of the textbook (p. 480-481).

In addition to the CBC, the Ministry of Education also passes on to teachers the National Curriculum Parameters (PCN's), which are cross-cutting themes that help teachers to work with students in their formation as conscious and participative citizens. One of the cross-cutting themes of the Curriculum Parameters is (Environment and Health). This volume refers appropriately to the subject of the environment, teaching the fundamentals, concepts, history and ways of working on the subject. The aim of the environmental parameter is to create an environmentally sustainable and socially just society. The environmental perspective consists of a way of seeing the world in which the interrelationships and interdependence of the various elements in the constitution and maintenance of life are highlighted (BRASIL, 2000).

It is therefore understood that, according to the PCN's, environmental issues constitute a cross-cutting theme which, although it is most clearly present in the disciplines of Science, History and Geography, also needs support from the areas of Mathematics, Portuguese Language, Physical Education and the Arts in order to be understood, given its interdisciplinary genesis.

> It is necessary for schools to deal with issues that affect students' lives and with which they are confronted in their daily lives. From this perspective, social issues relating to ethics, health, the environment, cultural plurality, sexual orientation and work and consumption are integrated into the educational proposal of the National Curriculum Parameters as Transversal Themes (BRASIL, 2000).

Contextualizing the content is an important tool. Teachers should develop questions with students that present different elements in order to make them think and act in a way that contributes to society. In this way, they acquire an ethical attitude towards life; they start to see things through their own eyes, reflect on their attitudes and start to respect others and the principles that govern life in their community. It is also important to say that: [...] "if we don't pay attention, if we don't conserve the environment, in the short or long term, we can deal with a lack of drinking water, food in quantity and quality, diseases and many natural disasters that would put our own lives at risk" (FLICK, 2010).

It is clear that teachers are not adequately prepared to work on environmental preservation. Regardless of their training, teachers must be adequately prepared to work on sustainability and environmental preservation, since it is everyone's duty to be participatory and take care of the biotic environment. Tozone-Reis, (2004, p. 151), mentions the training of educators:

> The training of environmental educators in higher education takes place in an unsystematic way, practically boiling down to three types of completely disconnected support: treatment of environmental themes in related subjects; optional environmental education subjects and educational-pedagogical training - in the different specialties - offered by education subjects in undergraduate courses. Although we consider the significant presence of activities with environmental themes identified in the courses investigated, the opportunities for reflection - offered to undergraduates - on the different theoretical references of environmental education are few from a quantitative point of view and superficial from a qualitative point of view. Not even the fundamentals of teaching practice are discussed, but they are incorporated and practiced invisibly.

It is necessary for children to learn values about the environment in which they live, to be aware of the importance of environmental preservation and to take these values into their homes and to those who live with them. The child as messenger is a common theme in the epistemology of this growing body of educators, poets, holistic theorists, traditional storytellers, spiritualists and elders. By shaping the individual as a child, we can guarantee a more effective conscious adult tomorrow (HUTCHISON, 2000).

At the Rio/92 International Conference, representatives from more than 170 countries recognized the importance of schools in building a sustainable society. Brazil has countless natural resources, one of the largest in the world, diverse ecosystems, abundant fauna and flora, and a large part of the world's fresh water. It is worrying, however, how Brazil's natural and cultural resources have been treated, despite the creation of the Law on Environmental Education, international concerns and targets set at UN Conferences, which have not yet been able to minimize the loss of resources in Brazil (BRASIL, 2000).

## 4 INFORMAL ENVIRONMENTAL EDUCATION

Informal education is when the individual acquires knowledge at home, at church, at work, with family and friends. Environmental education is very broad and can be carried out informally and formally. Formal education is where teachers can make students aware of the importance of environmental education (DIAS, 2004).

Within Law 9.795/99, (Section III - Non-Formal Environmental Education), Art. 13 "Non-formal environmental education is understood to be actions and practices aimed at raising the awareness of the community on environmental issues and its organization and participation in defending the quality of the environment. " (DIAS, 2004, p. 205).

Environmental education is very broad and can be carried out informally and formally. Formally, teachers can make students aware of the importance of environmental education, especially high school students, and they can communicate informally in their homes with family, friends and society in general (FLICK, 2010).

## 5 APPROACH TO ENVIRONMENTAL CONSERVATION IN THE SCHOOL CONTEXT

Every day we see advertisements on TV and on the internet that dictate what is right and mess with people's intellects. Sometimes people have never seen such an object, or have no interest in it, but the way it is expressed, the way it is taught, induces people to use and desire what is proposed. The school then enters this context, preparing individuals for society, individuals who seek sustainability, individuals who protect nature and the organisms that depend on it. Tozone-Reis (2004) reports on the environmental crisis caused by consumerism:

> The environmental crisis, which is a global "super-crisis", calls for a new approach to education, making environmental education a dimension of education. The different approaches to environmental education lead to the definition of different practices and objectives (TOZONE-REIS, 2004, p. 8)

Environmental education arose from the need to prepare people to conserve and protect nature, to seek quality of life without intervening in the environment. A society has the characteristics of those who live in it, in other words, the transformation of the individual transforms society, and this transformation can be achieved through education. The complexity of nature requires a systematic approach to its study, a work of definition and history, with various components seen as a whole, parts of a larger system, as well as their relationships and interactions with the means and regulation of life, and its aspects (BRASIL, 2000).

When questioning education professionals about the factors that hinder the inclusion of Environmental Education and how to overcome them, the answers are straightforward: We don't know how to do it. A simplistic answer to an essentially challenging problem: how to train educators. It is believed that educators lack a more solid methodology; it would be necessary to apply socio-pedagogical practices based on environmental themes; teacher and student must interact, bring innovations to the classroom that are capable of holding students' attention and developing their socio-environmental capacity. Classical methods that make the student feel like just a spectator of the teacher have no effect (FLICK, 2010).

Based on our understanding of environmental education, we understand that concern about this issue is not new. It is believed that education alone cannot solve environmental problems (although it does play a crucial role). Environmental issues must take into account the various systems involved.

> Environmental Education is an undefined and disorganized subject within the school curriculum, with isolated support. There is a tenuous project here and there, involving the students - often just to supplement their workload (FLICK, 2010, p. 15).

Environmental Education has a legal purpose in the project, but in practice it doesn't achieve its objectives, as the teaching staff in schools is not prepared to prepare the students. It is important to state that teaching and environmental education are two areas that are closely linked at all levels of society. In this respect, environmental education must consider the whole environment and pay close attention to environmental issues.

## 5.1 Environmental education and its institutional relationship

From the conceptualization of environmental education, we can see that concern with the subject is not new. This dimension permeates the educational universe in such a way as to give a better understanding of the process in which environmental education is embedded. It is worth emphasizing that education alone cannot solve environmental problems (but it does play a crucial role). Environmental issues must take into account the various systems involved (BRASIL, 2000).

According to Guimaraes (2005, p 44):

> Planning in EE starts from the local reality, but is inserted in the global reality, demonstrating the need to understand the specificity of each environment, as well as the link between the two realities.

According to the PCN's (BRASIL, 1997) "[...] the main function of working with the Environment theme is to contribute to the formation of aware citizens, able to decide and act in the socio-environmental reality in a way that is committed to life, to the well-being of each individual and of society". Therefore, more than just information and concepts are needed: schools must set out to work with proposals and attitudes, training and values, with the teaching and learning of skills and abilities.

It is acknowledged that formal education, by maintaining overloaded teaching hours and a curriculum organized in a disciplinary manner, does not provide interdisciplinary experiences, as health education and environmental education require. This contributes to local problems not being seen as issues to be tackled in the classroom, even though they are known by many teachers. One of the reasons for this is the lack of motivation on the part of teachers to deal with problems that are considered to be beneath them, especially if they are worked on in isolation (GRYNSZPAN, 1999).

The approach to the subject of environmental education falls more specifically within the disciplines of biology, science and geography. Even if there is a greater affinity between these disciplines in relation to the study of environmental education, it should extend to all the content, bringing it together with the other disciplines. Traditional disciplines alone are not enough to address environmental education.

environmental problems are so urgent that schools need to promote differentiated activities. In this problem-oriented educational process
all teachers must be involved, because environmental education is not just a subject like any other. Therefore, it is not the task of just one
teacher, but everyone's responsibility (MELLER, 1997; CALLAI, 1998
apud FARNESI; MELO. 2002)

Science and biology teachers use a wide variety of resources and ways to work on environmental education with their students, from theory to organizing events on commemorative dates related to the environment to ecological trips. Also according to the authors, the teachers reported that the process of transforming the students' awareness of environmental problems takes place through contact with nature, raising awareness and rescuing the link between man and the natural environment.

Brasil (1997) proposes the development of strategies to establish an effective culture of continuing teacher training, working with teachers centered on thematic axes. For environmental education, we could focus on the "quality of life" axis, which would focus on the concept of balance and its history as essential for this pedagogical contribution.

According to Queiroz et al (2013), it is of considerable importance that environmental education is included in the school routine in a stimulating, fun, critical and dynamic way so that it can develop environmental awareness in children, helping them to change their habits and thus preserve the biodiversity of the Brazilian Cerrado. It is interesting that the activities are geared towards the environment and that they can interact with teachers and students so that they achieve the proposed teaching-learning objectives and become thinking, sustainable citizens.

Lack of knowledge and failure to value the Cerrado are extremely important factors in its degradation. In this sense, environmental education needs to make everyone aware of the reality of local problems, so that in full awareness of them, teachers, students and the community can seek measures to help preserve the Cerrado. EE then raises awareness and transforms people's way of thinking and acting, restoring the interaction between man and nature and seeking innovative proposals to help preserve the cerrado.

EE is worked on in an interdisciplinary way in schools, increasing everyone's participation in the teaching-learning process. The focus is on raising awareness, increasing values and attitudes and forming citizens who are responsible for the environment. It also challenges the critical formation of students so that they can find solutions that result in effective practices and actions to guarantee the conservation and preservation of the Cerrado's biodiversity and natural resources. It should therefore be included in school life from the earliest years, and go beyond the confines of the school to be applied on an ongoing basis with the participation of the community and the family.

### 5.2 Environmental education at university

In view of the above, the production of knowledge must necessarily take into account the interrelationships between the natural and social environments, including the analysis of the determinants of the process, the role of the various actors involved and the forms of social organization that increase the power of alternative ideas for a new development, from a perspective that prioritizes a new development profile, with an emphasis on socio-environmental sustainability (JACOBI, 2003).

Law 9.795/99 itself, which establishes the National Environmental Education Policy, states that environmental education "must be present, in an articulated manner, at all levels and modalities of the educational process, both formal and non-formal" (art. 2). It is up to educational institutions, therefore, to "promote environmental education in a way that is integrated into the educational programs they develop" (BRASIL, Law 9.795/99, art. 3).

There has been much discussion about a better definition and the introduction of the environmental dimension into school education, with the objectives, principles, strategies and recommendations for the development of Environmental Education reflecting all the social, cultural, historical and political aspects that lead to the destruction of the environment in which we live and are part of (AZEVEDO, 1996).

Therefore, when education works from the concrete reality of the students involved, it enables and rescues the contextualized dimension of the content, since the students are challenged to overcome problematized everyday situations by perceiving themselves as being in the world and as the world, since "knowledge only exists in invention, in reinvention, in the restless, impatient, permanent search that men make in the world, as the world and with others" (FREIRE,1975),

Through research, the approach to environmental education has become a new and growing presence among the areas and lines of research within the field of education. In times when information plays an increasingly important role in the cyberspace: multimedia and the Internet, citizenship education represents the possibility of motivating and sensitizing people to transform the various forms of participation in defending the quality of life (RUSCHEINSKY, 2002a). As such, it is worth noting that environmental education is increasingly taking on a transformative role, in which the responsibility of individuals is an effective objective for promoting a new type of development - sustainable development. It is therefore understood that environmental education is a necessary condition for changing a situation of growing socio-environmental degradation (JACOBI, 2003).

Higuchi and Azevedo (2004, p. 08) state that:

> The seriousness of environmental problems means that measures to reduce the negative impacts on the natural environment and society must be

taken as quickly as our predatory action has progressed. This implies a huge environmental loss, due to the highly polluting potential and poor management of the waste generated, compromising the quality of the air, soil and, above all, surface and groundwater.

Educational work is undoubtedly one of the most urgent and necessary means of reversing this situation, as many environmental imbalances are currently linked to inappropriate human behavior, promoted by consumerist appeals that generate waste and the inappropriate use of natural resources. Therefore, it is probably through educational institutions that we will be able to change human habits and attitudes, forming ecologically aware individuals (AZEVEDO, 1996).

The accelerated and disorganized growth of Brazilian cities has led to a growing deterioration in living conditions, which imposes a necessary reflection and the confrontation of challenges to change ways of thinking and acting around emerging problems (SAUVE, 2005).

The university is a privileged space for the formation of citizens and the development of values that influence the acquisition of appropriate attitudes towards consumption and waste disposal, but educators are still unable to intervene in environmental education in a significant way, since they do not take into account the many facets of our relationship with the environment (SAUVE, 2005).

### 5.3 Conservation and sustainable use of natural resources

Every society wants to live in a better, more fraternal and ecological world. The conundrum is that everyone always expects this better world to impact on the other. For example, they choose to wait for a neighbor or friend to invite them to plant a tree or start selective waste collection, instead of taking the initiative (AVANZI, 2004).

Lately, there has been a lot of talk all over the planet about ecology, the environment and the sustainable management of renewable natural resources. However, only a small part of the population has sufficient knowledge to understand the dynamics and interrelationships that occur between the different ecosystems that exist in the world. It is necessary to work towards bringing information about the environment to all social strata, so that each individual can achieve an ecological awareness that is acceptable for reversing the frightening process of degradation that we are witnessing (SCHNEIDER, 1997).

Brazil is one of the richest countries in the world in terms of biodiversity, competing with Indonesia for the title of the biologically richest nation on our planet. Distinguished as it is, however, it rarely attracts attention for what it has; in particular, it is berated for what it is losing through deforestation; the conversion of natural landscapes into reforestation, soy plantations and pastures; and industrial and urban sprawl.

Brazil's biodiversity is home to between 15% and 20% of the world's plant, animal and microorganism species. Despite the growing international interest in Brazilian biodiversity, we know that this alone is not a sign of economic growth, nor of sustainable development. But it is hoped that the country, as well as being a respected exporter of raw materials, will become the protagonist of a new world economy, improved in the sustainable use of biodiversity and its derived resources, combined with value for the different productive sectors (RECH, 2004).

The constant increase in human population has caused major problems for the environment, because the larger the human population, the greater the consumption of food and natural resources. Excessive consumption leads to a large amount of solid waste, which has no specific destination, giving rise to garbage dumps and landfills that are unable to store it. Exaggerated demand for food, housing, energy, industrial production and transportation have a high environmental impact, which can be defined as a shock caused by human or even natural work that causes disharmony and imbalance in the environment (PANAROTTO, 2008).

Proof of the compatibility between the way natural resources are used and environmental protection was reaffirmed in studies on these populations carried out by the Nucleo de Pesquisas sobre Populagao Humana em Áreas Umidas Brasileiras (NUPAUB) at the University of Sao Paulo from 1987 onwards. The news of sustainable development applied to aid conservationist actions originated from a general discussion including the confrontation between the need for economic growth and the need to conserve natural resources (TEIXEIRA, 2005).

Discernment is about regulating the magnitude and extent of the exploitation of natural resources in such a way as to preserve the overall balance of ecosystems. The thesis thus goes beyond the sphere of individual private enterprises and into the global sphere of human action on the environment. The regulation of the levels of global human influence on the environment (CAVALCANTI, et al. 1994).

Uncontrolled urban growth, serious infrastructure deficiencies, poverty and an inert political structure, combined with the occupation of areas exposed to natural hazards, have created environments of intense vulnerability since the 1950s and a dampening of society's ability to respond to crises and emergencies, which can be confirmed by the increase in the frequency and magnitude of catastrophes in the region (ALMEIDA; PASCOALINO, 2008).

What is proposed is a paradigm shift, suitable for making humans less self-centered and more aware of our interactions with the natural environment, of which we are a part and with which we interact daily and intensely (FRANCO, 2009).

The real incompatibility lies between environmental preservation and the exaggeration of unsustainable consumption and the privileged accumulation of wealth. Sustainability serves to guarantee a better quality of life for all future generations, bringing together ecological and social interests and offering business opportunities for companies that can complete the lives of people and the world (PANAROTTO, 2008).

### 5.4 Recycling at school

School is the environment where students will try to learn to associate themselves fully with society. And it is in practice that students will have the chance to have this environmentally correct experience, so that they can become responsible citizens (NOGUEIRA, et al, 2013).

Among the serious environmental problems currently facing the planet (human society and nature), the issue of urban waste is a difficult one to solve, not least because waste generation is entirely linked to the population's power of consumption and, as such, along with economic growth comes an increase in waste production (SANTOS, 2008).

We always hear that: [...] "the school is defined as a means that prepares for life, therefore, the school as a social institution, establishes an ambiguous link with society, it is part of it and therefore works for it forming the individuals necessary for its maintenance". It is understood that the school's task is to ensure the development of society and, for this reason, it needs to create individuals who can produce transformations. "Therefore, the school cannot remain stuck in the past, in the old, in tradition; opening up the possibility for the emergence of a critical and innovative school" (DIAS, 2004, p. 02).

Lately, environmental issues have been discussed because of the need for change in relation to environmental degradation. In this context, education must be seen as an element in the transformation of societies, enabling the development of a new morality that is different from that which guides a consumer society. Environmental education should be worked on in a unified, successive and permanent way at all levels and in all forms of education, as stated in the National Curriculum Parameters, which serve as contributions to pedagogical practice, in the knowledge that waste is one of the biggest problems directly affecting all social and environmental issues (FELIX, 2007).

Environmental education is seen as a way of actively changing reality and the conditions for quality of life, through awareness-raising that takes place through reflective social practice and dialog. This process is essential for forming sustainable societies, i.e. societies that are guided to face the challenges of today, guaranteeing quality of life for this and future generations. Therefore, environmental education should be covered in its broadest sense, aimed at training people to exercise responsible and conscious citizenship, and to have a broader perception of the environments in which they live (LOUREIRO, 2006).

Today, there is a clear need for the school to interfere directly in the formation of individuals capable of relating to the environment, always seeking to achieve the knowledge, values, attitudes, commitments and skills needed to protect and improve the environment. To this end, it is important to address environmental issues and an integrated view of the world, as well as the condition of life, right from early childhood education (RODRIGUES, 2010). This vision is also vital and decisive for the topic of recycling.

In this context, tackling the issue of waste management and selective collection through the lens of citizenship and environmental education necessarily involves seeking political participation in order to overcome everyday shortcomings. For this to happen, educational methods also need to overcome the dichotomy between the individual and the whole, acting on the network of meanings that is culture itself, and reinforcing its role as a support and language for a perception of shared common interests, which are the essence of citizenship (ZANETI; SA, 2015).

Environmental education is a participatory process, in which the student becomes a central member of the intended teaching/learning process, actively participating in the assessment of environmental problems in search of solutions, including awareness of recycling. The affinity between environmental education in schools and recycling is of fundamental importance, raising awareness of citizenship and preparing them to participate actively in today's demanding society (NOGUEIRA, et al, 2013).

When asked about the destination of their garbage, they usually reply that a truck takes it away for recycling, but most of the time they don't know what recycling is or the right material to recycle. Generally, the only form of waste collection that the students know about is selective collection, because the truck goes around their houses and buildings, but they don't notice the difference between the selective collection truck and the other ordinary garbage truck, they just reproduce what they hear people say about the recycling truck. Through this perception, it is necessary to delve deeper into the subject of waste and look for ways to include the subject of recycling within the school context (RODRIGUES, 2010).

According to Santos (2008), it is important to show that waste recycling is not just another school project, but the support that everyone can give to improve the planet's environmental conditions. In this way, it's important that sorting waste also becomes a habit for everyone, because then, when recycling projects are implemented in schools, everyone will be involved in collecting the material, regardless of their position in the school or, in the case of teachers, the subject they teach.

Still according to Santos, (2008) another important point of discussion that waste recycling at school can provide is the question of society's current consumption pattern and its relationship with the immense generation of waste. Differentiating necessary consumer goods from superfluous consumer goods is imperative if we are to form conscious citizens and perhaps even think about sustainable consumption.

### 5.5 Sustainability and recycling in question

Since the publication of Agenda 21 at the UN Conference on Environment and Development (United Nations Educational, Scientific and Cultural Organization) held in Brazil in 1992, there has been an innovation in people's way of thinking, and consumption has been discussed worldwide: this is sustainable consumption, the suggestion of which is to establish a way of consuming that is capable of satisfying the needs of current and future generations (BIERWAGEN, 2007).

The buzzword that is sustainable development instigates a set of paradigms for the use of resources that aim to meet human needs.

Recycling is the driving force behind the sustainability process, comprising various technical, economic and social aspects of the relationship between man and the environment. Understanding the importance of recycling is the first step, but knowing how to practice it is the biggest challenge. Contrary to what many people think, the cost/benefit ratio of a well-managed recycling project can have surprising positive effects, and we can probably list a few cases throughout Brazil (RIBAS, 2007).

In this new century, the discussion of sustainability is playing a leading role in reflecting on the dimensions of the increase and the alternatives that are emerging. The socio-environmental representation that distinguishes contemporary societies shows that the impact of humans on the environment has increasingly complex consequences, both in quantitative and qualitative terms (JACOBI, 2003).

According to Ribas, (2007, p. 3) "the secret to the success of environmental preservation projects lies in their economic sustainability. The businesses that materialize in the various stages involved are the guarantee for the continuity and improvement of such initiatives".

Sustainable development is a dazzling possibility and its construction is possible because the current crisis of the patterns that drive industrialist progress authorizes the audacity to think of another way of human development, without harming the environment (BURSZTYN, 2001).

The issue of environmental management has been gaining a significant foothold in the business world. The increase in environmental awareness in different sectors of world society is beginning to involve the business sector as well. Therefore, it cannot be said that all sectors
Businesses are already aware of the importance of managing natural resources responsibly. Any business that does not try to adapt its activities to the judgment of sustainable development is about to lose competitiveness in the short or medium term (KRAEMER, 2005).

The issue of sustainability confronts the "risk society" model. This highlights the need to correct social methods based on strengthening the right to access information and environmental education in an integrative way (JACOBI, 2003).

One of the methods is the conscious consumption of products that generate less waste and also a society with new thinking around the subject of recycling.

In order for environmental education to effectively achieve sustainability, it requires a continuous learning process, enhanced by respect for all forms of life, guaranteeing values and many actions that contribute to the social development of man and the preservation of the environment (MATTOS, et al, 2010).

In order to contribute to sustainable development, the political and economic classes and societies are already looking for sustainable sources of energy. The demand for energy is growing, but the resources available are limited. In addition, the use of fossil fuels such as oil and coal causes considerable damage to the climate and the environment. Materials such as manure and industrial and residential waste provide electricity, heat and fuel.

Even with all the popularity of the term sustainable development, it is crucial to make it a reality, but if we want a better world and quality of life for our descendants, it is up to us to seek improvements in order to preserve the environment, and for this it is essential to use natural resources in a sustainable way (FERNANDES, 2009).

Raising awareness that recycling can provide major economic gains will certainly help speed up the sustainability process. The study of knowledge from other countries can also contribute, serving as a direction for Brazil (RIBAS, 2007).

## 6 ETHICS, THE ROLE OF THE CONSUMER AND THE FUTURE

The fate of all species depends on the stabilization of the environment, because without a homeostatic relationship between human beings and nature, there is no way to guarantee a good quality of life today and for future generations of the entire ecosystem (GOMES, 2006).

According to the aforementioned author, we understand the need to look for a new ethic, driven by a longing for mutual belonging between all beings. Ethics has always been absorbed with questions of human existence, but now it must turn, above all, to its interrelationship with the planet, an ethics focused on a balanced relationship between nature and human beings. According to Carvalho (2003 p. 2):

> No one is born a citizen, but becomes one through education. Because education actualizes the possible and natural dedication of men to community or social life. It is a process that began in the dawn of humanity and that is activated through the knowledge and conquest of human rights, not as something ready, finished, but as something that is built. Just like ethics, citizenship is today a major issue, both in education and in the family and organizations, for improving the way of life of an entire society.

In other words, there is a belief in the ability of accumulated knowledge to be directed towards solutions in the fields of exact and biological sciences. In the cultural, political and other human sciences, exceptionally, history has shown the narrow limits of diplomacy, often stretched by wars, when antagonisms were much smaller than those that the future holds. Recent years have shown a new element that is mitigating conflicts and expanding the scope of diplomacy: the globalization of the economy. But it's too early to say that peace is on the horizon just because trade has created a protective network of interests, sustainability (ZULAUF, 2000).

Environmental education, therefore, has the task of contributing to the process of changing today's society into a sustainable one, centered on the responsible exercise of citizenship, which considers nature as a common good, takes into account the regenerative capacity of material resources, promotes the equitable distribution of the wealth generated and contributes to decent living conditions for current and future generations (PELICIONI, 1998). Education plays an important role in making society aware of the importance of the environment.

## 7 THE EDUCATOR AND THE TEACHING-LEARNING PROCESS ABOUT RECYCLING

It seems that by working on the subject of environmental preservation in an appropriate and reflective way in schools, taking into account that it should not be limited to the school environment and extending it to the local community, we can contribute to the recovery and preservation of natural resources and improve the population's quality of life. Everyone has the right to a healthy environment, without pollution of any kind (SALES, 2011).

It is essential to develop environmental education activities in order to motivate students to participate more as citizens in the municipal cleaning system, showing them the environmental, economic and social consequences of simple, everyday acts such as the correct packaging of waste and observance of collection times (EIGENHEER, 2009).

It is in this context that the educator appears to be of essential importance so that the subject of waste can be debated in a profound and real way, in order to awaken in the student (in the child, in the citizen) the awareness of thinking about the environmental issue not as an educational fad, but as a practice and a necessity to maintain, and to possess, an environmentally healthy planet for current generations. Teachers and educators have the task of broadening this theme beyond the didactic-pedagogical limits of each subject, permeating all areas of knowledge in an interdisciplinary way, involving the whole school and then the whole school community (SANTOS, 2008).

In order to work with the context of recycling in school, we need a didactic basis and, above all, methodologies that are committed to this aspect. In this way, in the teaching-learning method, students and teachers come face to face, and both take part in the same experience that will make the first men spiritually adults. The teacher gives the floor to the pupil so that he or she can pronounce the essentials; it follows that the main thing is not teaching, but learning. It is, above all, a personal and human relationship, the meaning of which varies according to the age and personality of those involved (BRASIL, 2001).

According to Gomes (2003):

> By learning how to build and use varied and easily accessible teaching resources, teachers will be able to better guide their students. The teacher's duty

> is to mediate the relationship between students and knowledge, organizing the group and prioritizing teaching activities that can be meaningful for learning, according to the reality they work in and the profile of their students. We mustn't forget that knowledge doesn't just happen through written or spoken texts, but also through smelling, touching, tasting, etc. (GOMES, 2003, p. 15).

It is necessary to prepare students so that they can construct knowledge and use it in their daily lives, making them critical, democratic and participatory citizens (GONQALVES, 2000).

Therefore, interest in the student's self-fulfillment provides the conditions for essential learning, i.e. learning that leads to the satisfaction of the student's objectives, instigating creativity, imagination, self-awareness as a being capable of making their own choices and taking responsibility for them, becoming a good chooser and opinion-former (BRASIL, 2001).

In turn, the teacher has personal needs and objectives represented by his personal history, which in turn influences his choice of objectives for the student. His frame of reference represents standards derived from personal education and professional experience with which he evaluates his actions and the actions of his students, as well as the relevant features of the environment. These evaluations lead to the choice of new actions. Feedback is the essential element that allows the teacher to evaluate him/herself and the student's actions, so there is an interdependence between the actions of the teacher and the student (BRASIL, 2001).

According to Talamoni; Sampaio (2003), a major problem in the education today is the lack of motivation and involvement of students in learning processes. Especially since for children it is adolescents, the understanding of reality is based more on concrete aspects than on abstract ones. This is closely linked to the idea of recycling.

It should also be realized that students need to be involved in order to assess the magnitude of what is being taught, whether or not they enjoy the activities, and thus form individuals who can use the knowledge to act consciously in their reality. According to Gongalves (2000, p. 23):

> [...] the human brain and the rest of the body constitute an inseparable organism, forming an integrated whole through mutually interactive biochemical and neurological regulatory circuits. He argues that the organism interacts with the environment as a whole: the interaction is not exclusively that of the body or the brain and that the physiological operations that are called mind derive from

> this structural and functional whole and not just from the brain: mental phenomena can only be understood in the context of an organism interacting with its environment.

As explained by Talamoni; Sampaio (2003), this means looking for new teaching perspectives that consider students as complex individuals and that learning depends as much on the power of logical reasoning as on elements considered subjective, such as feelings linked to pleasure, satisfaction, beliefs and motivation to learn.

### 7.1 The importance of plastic recycling in schools

It is important to promote environmental education through interdisciplinarity. In the different and diverse subjects taught at school, it is necessary to use the syllabus and direct it so that students can solve a visible problem or understand a phenomenon from different points of view, seeking to build a critical and participatory citizen. It is in this context that the educator appears to be of essential importance so that the subject of waste can be debated in a profound and real way, in order to awaken it in the student (in the child, in the citizen) (MATTOS, et al, 2010).

The subject of polymers is one of the topics required in chemistry teaching programs in particular, but it is also a theme in environmental education, but most of the time it is centered on scientific concepts, without implication in real situations. By addressing the problem of waste in society, the subject becomes more exciting for the student, facilitating learning (SANTA MARIA, et al, 2003).

In most cases, recycling is only possible industrially, but reuse can be done by hand, for example by turning PET bottles into sofas or brooms (NASCIMENTO, et al, 2005).

The most recycled plastic today is the PET bottle. The most sought-after plastic for recycling, PET is discovering new vocations at a rapid pace. Consolidated in various market segments, including clothing, recycled PET is now making headway in the manufacture of paints, pipes, flooring and coatings (SINO, 2001).

The importance of environmental issues at school is mostly focused on the practice of composting, setting up a selective collection system and workshops to make objects from reused and recycled materials (MEIRA, et al, 2009).

However, in order for waste recycling at school not to become an empty school project, it needs to be carefully planned, so knowledge of this subject is indispensable for success in a school endeavor (SANTOS, 2008).

It is important to discuss with students the biggest problem with plastics, which is their resistance to biodegradation, mainly due to their chemical nature. Point out that they are resistant to radiation, heat, air and water, remaining in the environment and retaining their physical properties, and that it would take 100 to 150 or even more for these polymers to degrade in nature (NASCIMENTO, et al, 2005).

Special care must be taken with plastic because of the great dependence that the current consumerist way of life has on this petroleum resin (polymer). The reuse of packaging, together with reduction and recycling, make up the trilogy proposed by Agenda 21 for dealing with waste in order to reduce its quantity. Reuse suggests using the packaging of a product for another function with little or no alteration to its original structure. However, nowadays, due to the great specificity of the products, the productive sector, creator and owner of the packaging, has given it little reuse (SANTOS, 2008).

In this way, the reuse of packaging affects ordinary citizens who, through campaigns promoted by the television media, newspaper and magazine articles and other means of communication, are encouraged to reuse all types of packaging in their homes (SANTOS, 2008).

Recycling is understood to be a technique that aims to take waste and reuse it in the production cycle from which it came. It is the result of a series of activities whereby materials that would have become waste, or are in the waste stream, are diverted, collected, separated and processed to be reused as raw materials in the industry for new products (SANTA MARIA, et al, 2003).

Recycling at school trains sensitive, aware and multiplying citizens, even though it is known that transforming habits and behavior takes time to develop, and that it is also considered essential to form partnerships to better encourage the community and obtain better results, with a more intense reach (FELIX, 2007).

Understanding this issue, the realization of projects that work on cross-cutting themes at school sensitizes students to learn how to act in the face of environmental issues, since they begin to recycle the waste also generated in their homes, since the results of recycling are significant in the environmental field, as well as in the economic and social fields. In addition to learning about the problem of waste and the need to act on these problems, students can be very enthusiastic when they realize the interdisciplinary nature of recycling projects, with the help of their teachers (JESUS, et al, 2011).

It is important to note that the students, after the environmental education action for recycling, recognize how important it is to carry out projects and that this type of work contributes significantly to minimizing the problems caused by waste (JESUS, et al, 2011).

According to Andrade, (2000, p 45):

> The school needs to position itself in an implementation process that is not hierarchical, aggressive, competitive or exclusivist, but that is carried out on the basis of cooperation, participation and the generation of autonomy for all the students and staff involved. Projects imposed by small groups or isolated activities, managed by just a few individuals from the school community, such as a selective waste collection project in which the only participation of the students is to throw the waste in separate bins, involving just one coordinating teacher, are not capable of producing the change in mentality necessary for the attitude of reducing consumption, reusing and recycling solid waste to become established and transcend beyond the school environment.

It is therefore up to families, schools, universities, public and private institutions to play their part and act responsibly, socializing environmental issues so that the community can actively share and benefit.

> Lectures serve to raise awareness among the school population through environmental education, arousing concern about consumerism, and helping them to learn about the challenge of recognizing and collecting plastic, so that waste can be separated correctly in households, effectively contributing to the necessary attitudes and transformations indicated and defended by Environmental Education, they help to recognize that it is possible to promote the reuse of plastic in their daily lives, making better use of the same packaging more often, or even reducing the circulation of some plastics, such as plastic bags, by replacing them with returnable bags made from cotton (MATTOS, et al, 2010).

It is necessary to train teachers through discussion forums on environmental education, as the information available in textbooks is precarious, and teachers often don't have access to textbooks. Selective waste collection in schools is fundamental, but this action must be constantly accompanied by a debate about the depletion of natural resources, which cannot be wasted (LOPES, CADEMARTORI, 2007).

And when it comes to waste, we need to understand that it can be consumed again through recycling or even reuse. It is necessary for students to realize that selling waste can also be an alternative for creating jobs, as well as being economically important, but also socially important, because the environment is everyone's property (LOPES, CADEMARTORI, 2007).

However, all of these actions will only be successful if the recyclable solid waste, such as plastic, reaches its final destination for recycling, either through selective waste collection or through the environmental awareness of the students who have asked for the waste to be collected, recognized, separated and sent to a cooperative or collectors, requesting solidarity and responsible action with the people who need this material to earn an income and being sure that the waste will go back into the production process of the industries, reducing the impact they generate by taking their raw materials from nature (MATTOS, et al, 2010).

As presented in the chapter, and in line with the authors cited, it can be seen that school is the environment where students will try to learn to associate themselves fully with society, in other words, it is at school that students will be able to complement their education, including with regard to recycling and environmental education.

It is well known that discussions and reflections on the environment have been in vogue recently, due to the need for transformations in relation to environmental degradation. As Felix (2007), Loureiro (2006), Zaneti; Sa, (2015), Santos (2008) and other authors have shown, this reinforces the idea that the educational process is linked to environmental reflections, especially in relation to waste. Therefore, environmental issues should be remembered as elements for transforming societies, enabling the development of a new morality that is different from the one that guides a consumer society.

In short, reflections on environmental education in the context of plastic recycling lead us to believe that today this topic is seen as a possibility for the active transformation of reality and quality of life conditions. As mentioned, it is in this context that the educator appears to be of essential importance so that the subject of "waste" can be debated in a profound and real way, in order to awaken interest in the subject in the target of awareness-raising.

## 8 ENVIRONMENTAL EDUCATION AND WASTEWATER TREATMENT

Throughout history, man has transformed the environment, created new cultures, established economic relationships and devised new models of communication with nature and all the things in it. These socio-economic and environmental relationships lead man to make decisions aimed at objectives that should be common to everyone in society, i.e. cultural growth, quality of life and environmental balance (BRASIL, 2001).

For a long time, man has linked effluents to ill-health, so throwing them away has always been a common practice, and the most usual places are rivers, lakes and ponds. But in recent decades, environmental problems have become increasingly critical and frequent, especially due to the large population growth and the increase in industrial activity. This has meant that the problems have reached flagellating proportions, which can be seen through changes in the quality of the soil, air and water (BEZERRA; REIS, 2009).

The fact is that the discharge of untreated effluents causes major damage to the environment and to society as a whole. In the waste there is a large load of waste which, after being washed into the water, generates an enormous growth of aerobic bacteria (PIMENTA et al., 2002).

This ends up causing major impacts on the aerobic aquatic environment and leads to the spread of diseases from contaminated water caused by ecological imbalance (DURAN, et al., 1997). The contamination of natural waters has certainly been one of the major problems facing contemporary society.

Thus, Environmental Education must be at the service of encouraging society to participate effectively in solving environmental problems, because according to Gongalves (2004), only the articulation between the physical environment, social relations and human subjectivity can try to solve environmental issues.

> [...] processes through which the individual and the community build social values, knowledge, attitudes, active interest and skills aimed at conserving the environment, essential for a healthy quality of life, which must be present, in an articulated manner, at all levels and sectors of the educational process (motivator) (CARVALHO, et al., 2000, p.3)

It is worth remembering that there are various human activities that can generate impacts of lesser or greater magnitude on the environment in which people live, making them catastrophic for life.

The absence of sewage treatment and appropriate sanitation conditions, for example, can contribute to the spread of various parasitic and infectious diseases, as well as deteriorating water quality. Proper sewage disposal is fundamental to protecting public health. Sewage produced by humans generates around fifty species of infections and can be transmitted from a sick individual to a healthy one in various ways. Sewage can infect water, food, household materials, hands, soil or even be carried by mosquitoes, cockroaches and rodents, thus causing new infections (UNICAMP, 2013).

Investing in sewage treatment can be a leap forward in terms of the infrastructure required to protect the environment and can improve people's quality of life. Therefore, the collection, treatment and environmentally appropriate disposal of effluents are essential. Right,

> [...] for every R$1.00 (one real) invested in the sanitation sector, R$4.00 (four reals) is saved in the area of curative medicine. Approximately fifteen (15) children between the ages of 0 and 4 die every day in Brazil due to a lack of basic sanitation, more precisely, a lack of sanitary sewage (CARVALHO, et al., 2000, p.31).

This means that proper sewage treatment is fundamental to protecting public health and that the absence of sewage treatment and proper sanitation conditions can contribute to the spread of various diseases.

Epidemics such as typhoid fever, cholera, dysentery, infectious hepatitis, and numerous cases of worms and some diseases that can be transmitted by improper sewage disposal are responsible for high mortality rates in underdeveloped countries. The association of these diseases with malnutrition is often fatal, especially for children. An increase in life expectancy and a reduction in the prevalence of worms, which are not usually fatal but do wear the individual down, can only be achieved through proper sewage treatment (UNICAMP, 2013).

The consequences of water pollution:

> [...] impacts on the quality of life of the population; transmission of diseases; damage to water uses; silting; eutrophication; aggravation of water

scarcity problems; increase in the cost of water treatment; ecological imbalances; degradation of the landscape (SANTOS, 2013, p. 1).

Another important reason for treating sewage is to preserve the environment. The substances present in sewage have an unhealthy effect on bodies of water: organic matter can cause a reduction in the concentration of diluted oxygen, which can lead to the death of fish and other aquatic organisms, as well as darkening the water and giving off odors; detergents present in sewage can cause foaming in places where the liquid is agitated; pesticides cause the death of fish and various animals. There is also the prospect of eutrophication due to the presence of nutrients, which cause algae to grow rapidly (UNICAMP, 2013).

The importance of effluent treatment is also reflected in legal aspects such as:

- Law 11.445/2007 - National Basic Sanitation Policy, which establishes national guidelines for basic sanitation (BRASIL, 2007).

- COPAM/CERH Joint Normative Decision 01 - Effluent discharge for the state of Minas Gerais, which provides for the classification of bodies of water and environmental guidelines for their classification, and also establishes the conditions and standards for effluent discharge (MINAS GERAIS, 2008).

- NBR 12.207 - Projetos de Estagoes de Tratamento de Esgotos this standard establishes the conditions required for working out the sanitary hydraulic design of sanitary sewage interceptors, analyzed exclusively by the planning and development of the sanitary sewage system (ABNT, 1992).

- CONAMA Resolution 430/2011 - Effluent treatment (Law Federal), which lays down conditions and standards for effluent disposal (BRASIL, 2011).

It is worth noting that Environmental Education, due to its interdisciplinary nature, is an important instrument for the development and practice of policies aimed at improving the quality of life, especially in large urban centers. Therefore, environmental education is a subject that is already present in the daily lives of many different segments of society, and is increasingly gaining ground in educational institutions, organizations and industries. However, presenting alternatives to promote sustainable development is a major challenge for society, even for the government, which is creating programs to encourage the inclusion of environmental education in the actions of organizations and entities linked to the environment (CARVALHO, et al., 2000).

In view of the above, it is extremely important that environmental education is guided by engaging leaders who are persevering and resigned, supportive and open to innovation. This is because "[...] educating is guaranteeing the formation of mature, adult collaborators who are attentive to problems and solutions". An example in Brazil is the Ethos Institute, which works to publish corporate experiences of socio-environmental actions with solid results, and is an administrator that guides the action of the Environmental Education process carried out by organizations (BEZERRA; REIS, 2009).

Sewage treatment is very important due to the large amount of waste that is produced in cities and which cannot be deposited anywhere, as it would prevent nature from decomposing quickly, causing different diseases. For this reason, treatment plants are needed to speed up the process and return water to the environment with smaller amounts of pollutants (BEZERRA; REIS, 2009).

People, institutions and society in general need to be made aware of the need to encourage changes in values and attitudes that could trigger a new development process (BRASIL. 2000).

Human beings have no respect for the environment in which they live, they are changing planet Earth and the consequences of this act are enormous. It is clear that environmental catastrophes are most often linked to human intervention, and as a result, millions of people are killed or left homeless. Hutchison (2000, p. 24) states that:

> Many diseases that arise as a result of deforestation include: malaria, cholera, leishmaniasis, chagas disease, among others; diseases that kill thousands of people every day. Social imbalance is also linked to the exploitation of natural resources. Some countries have more resources than others and use them to increase their GNP (Gross National Product). It is known that degrading the environment can lead to the end of life on the planet, and yet nothing is changed. People criticize those who don't respect nature, but they don't set an example, they don't teach their children about the importance of the biotic environment. Proposals for sustainability are created, but not supported by the population; the percentage of people who contribute to changing habits is negligible (HUTCHISON, 2000, p. 24).

EE is an area of education whose aim is to disseminate knowledge about the environment in order to help preserve it and use its resources sustainably, and a way of changing society's habits. "The role of environmental education in this context is becoming more urgent. We need to offer more training. Education still 'trains' students to ignore the ecological consequences of their actions" (DIAS, 2004, p. 16). Ecological awareness is needed on the part of society as a whole, because environmental impacts are the result of man's actions, which are the main cause of ecological imbalance today, such as deforestation, hunting and fishing without control, allied to urbanization in areas of woods and forests.

Brazil is a country of countless riches, where flora and fauna are abundant and very diverse, and water resources are also privileged and stand out from other countries. Brazil began to lose its resources in the 1500s, when the Portuguese docked their ships on the Brazilian hillsides. Pero Vaz de Caminha wrote to the king of Portugal, Manuel I, that the discovered land was full of exuberance, with rich flora and fauna (DIAS, 2004).

Many efforts have been made by governmental and non-governmental organizations to promote environmental education. In basic education, for example, the National Curriculum Parameters (PCN's) on environmental education were drawn up; a document approved by the National Education Council (CNE). The PCN's point to the need for interdisciplinarity in the study of the cross-cutting theme (BRASIL, 2000).

## 9 ENVIRONMENTAL CRISIS

"The environmental crisis has a direct and profound relationship with social and economic development, and the category of analysis of the relationship between man and nature meets this perspective" (TOZONI-REIS, 2004, p. 20). It is complicated to introduce sustainability and environmental preservation policies in an age of consumerism, when individuals and governments are only concerned with financial capital and leave nature issues to one side.

"In any case, natural resources and the environment itself have become a priority, one of the most important components for the political and economic planning of governments" (BRASIL, 2000, p.22). Without resources
Without production, there is no capital, and what matters most to the rulers and the capture of capital are large-scale profits. If we look at the way people live today, we can see an increase in the population's consumption; globalization and advanced technology have turned human beings into compulsive shopping puppets.

Environmental conservation is a subject that has been gaining prominence in the media over time, and is currently the subject of much discussion in schools, governments, social groups, etc. As the population has grown, so has the exploitation of natural resources, people's consumption and territorial expansion. Tensions and conflicts then arose over the use of space and resources as a result of the technology available. Technology and industrialization have evolved so rapidly, causing undesirable and sometimes irreversible consequences, that the exploitation of nature has become intense. Non-renewable resources, such as oil, threaten to become scarce (BRASIL, 2000).

Cascino (2003) points out that:

> When the First World War broke out in 1914, there had been no major war for over a century. After the Second World War, the age of empires was over. After the explosion of the atomic bomb over Hiroshima and Nagasaki, humanity experienced another era. The organization of nation-states, cultures and technologies underwent profound changes. This situation would define the new world scenario; the change in mentality, experienced by all territories, would open the way for other revolutions (CASCINO, 2003, p. 26).

The planet is constantly changing, undergoing various transformations that can be natural or caused by anthropic action. Man has the capacity to modify, destroy and build up geographical space in relatively short periods of time. The consequences of human actions are great and are already being felt. It is now up to each individual to review their actions and change their human-nature relationship (CASCINO, 2003).

In Brazil, the idea of environmental preservation gained recognition in the 1970s, when the realization dawned that the degradation of the planet could have irreversible and catastrophic effects. There is a great shortage of water in many places on the planet, increasingly abundant diseases, crowded and increasingly industrialized cities, animals that are extinct or have no habitat in which to survive. A series of facts have led to various questions about the degradation of non-renewable resources. It's time to reflect on yesterday and now, on the nature we once had and now no longer have, on the relationship between climate problems and our intervention in nature. The energy spent by man, energy that is taken from nature through countless processes, is far more than can be replaced. Much more is spent than the Earth has to offer. Another planet would be needed to meet all the needs of human beings, but unfortunately this is not available (CUNHA; GUERA, 2008).

Another problem is the large amount of solid waste produced in Brazil, which goes to inappropriate places, polluting rivers, soil, air and the living organisms around it. It should be borne in mind that everything is discarded and thrown away, and realize that this "outside" is planet Earth, that everything that is discarded stays in the same environment that living organisms inhabit, garbage that accumulates and is bad for the environment and for health. "The decomposition of the organic material present in waste produces leachate, which is a dark-colored liquid with an unpleasant smell that can contaminate water and soil". (MAZZINI, 2012, p. 16).

Looking at the figures relating to waste in Brazil, you can see the seriousness of the risk that Brazilians are facing: - there are 400 cooperatives and around 1 million waste pickers; - Brazil loses R$8 billion a year by burying recyclable waste; - only 13% of waste is recycled and, unfortunately, 87% of waste goes to landfills or garbage dumps; - good waste management can reduce Brazil's carbon emissions by up to 5%; in Brazil, only 443 municipalities operate selective collection programs (MAZZINI, 2012).

Change is needed and it must be done urgently; we cannot turn a blind eye to the facts, to the events that are already directly and indirectly affecting all organisms. "Such a change will require a total reversal of current political and social thinking; government and communities around the world must realize that biological diversity is of the utmost value [...]." Environmental projects must be taken seriously, and people must think of their actions as affecting the whole. It is believed that it is important to teach people ways of preserving the environment, how to help with this, ways of recycling, socially ecological methods, changing society and individuals' social relationships. "The development of environmental education proposals leads to reflection on environmental problems and their relationship with education" (TOZONI-REIS, 2004, p. 8).

"Some government projects will only acquire meaning with individual and collective participation, in order to reach the causes of the problems they are intended to tackle. " (RUSCHEINSKY, 2002 p.12). Unfortunately, government support has not been very successful, and environmental meetings such as Rio+20 have not achieved their goals with preservation support.

Despite the international conferences and the implementation of the Environmental Education Law, improvements are still needed in the way environmental education is dealt with. Since environmental education is included as a cross-cutting theme in the school curriculum, it lacks credibility among educators and teachers need to be trained, not just science teachers, but everyone. Many think that it is the Science or Biology teacher's obligation to deal with the subject of environmental education in the classroom, but if the problem of environmental degradation is an international factor that involves all individuals, it is not fair to place the responsibility for environmental education solely on the Science/Biology teacher. If it's everyone's responsibility, then everyone should learn to deal with it, and every teacher, regardless of their field, should have the knowledge and skills to help their students understand these issues.

Without proper preparation, teachers will not know how to explore issues related to environmental education with their students, because it is during school practice that teachers must make students aware of the importance of these studies, develop sustainable habits and attitudes that seek environmental preservation.

It can be seen that the presuppositions of environmental education encourage the formation of citizens who are able to act in socio-environmental reality to improve their lives and the lives of other people in local and global society. Thus, according to what has been presented, it was found that EE today emphasizes curricular proposals, promoting the formation of citizenship in both school and non-school (informal) ways, as well as observing that the contextualization of significant aspects improves the "paths" for the formation of knowledge in Environmental Education.

## 10 ENVIRONMENTAL EDUCATION AS A MEANS OF CONSERVING THE CERRADO

From the point of view of biodiversity, the Brazilian Cerrado is recognized as the richest savannah in the world, home to diverse ecosystems. Considered to be the world's last agricultural frontier, the Cerrado has been suffering from the advances of agriculture and livestock farming and the indiscriminate use of agrotoxics, especially in recent decades, causing major problems for the biome, from habitat fragmentation, biodiversity reduction, drying out and contamination of water resources and soils, among others.

Environmental Education as a subsidy for the conservation of the Cerrado, seeks to integrate man with nature, raising awareness, adding values and attitudes, promoting skills that allow him to use natural resources in a conscious way, promoting economic development together with environmental protection.

The Cerrado has a great diversity of habitats and species, most of which are endemic, i.e. they are not found in other biomes, making them specific. A third of Brazil's biodiversity is also found in this biome, corresponding to 5% of the world's fauna and flora, yet it is considered one of the world's *hotspots*, which are areas that have been suffering major impacts and are in urgent need of conservation (SANTOS, 2010).

According to Fernandes; Pessoa (2011), the great biodiversity of the Cerrado is related to its large territorial extension, being considered the second largest biome in Brazil, containing diverse types of ecosystems that are constantly being modified by man, which represents a great loss from the genetic point of view due to the lack of studies and knowledge of these habitats.

The Cerrado has a variety of phytophysiognomies, including clean grassland, dirty grassland with the presence of herbaceous shrubs, Cerrado grassland with a predominance of shrubby plants, cerradao with shrubby vegetation 15 to 20 meters high and veredas where you can find the Buritis and the springs of water sources. The great biodiversity of the Cerrado ecosystem is also related to its various phytophysiognomies (MACHADO, et al, 2004).

According to Guimaraes (2005), Environmental Education (EE) plays an important role in integrating man with nature. Still according to the author, a relationship of harmony and awareness provides a dynamic balance in nature, through new knowledge, values and attitudes, the participation of students and educators and the whole of society as participating citizens in the process of transforming the current reality of environmental degradation.

It is believed that environmental education will help to form citizens who are aware of, committed to and responsible for development that mitigates impacts on the environment, respecting all forms of life in it.

### 10.1 Characterization of the Cerrado

The Cerrado is Brazil's second largest biome and is recognized as the richest savannah in the world. Its diverse ecosystems are home to a flora of more than 11,000 native plant species, 4,400 of which are endemic (KLEIN, 2002).

The Cerrado has two well-characterized seasons: dry winter and rainy summer. With tropical savannah soil, deficient in nutrients and rich in iron and aluminum, it is home to xerophytic plants, including sparse shrubs and grasses, and the cerradao, a denser type of vegetation, of forest formation (SANO; et al 2008)

The Cerrado or Cerrado grassland is, as already mentioned, Brazil's second largest plant formation. It originally extended over an area of 2 million $km^2$ , covering ten states in Central Brazil. Today, only about 20% of this total remains. Typical of tropical regions, due to the advance of anthropic intervention (PAULA, 2011).

> The transformations that have taken place in the Cerrado have also caused major environmental damage, such as habitat fragmentation, extinction of biodiversity, invasion of exotic species, soil erosion, pollution of aquifers, degradation of ecosystems, changes in fire regimes, imbalances in the carbon cycle and possibly regional climate change (KLINK; MACHADO, 2005).

The Cerrado biome, being located in a central region of the country and bordering the Atlantic Forest, Amazon, Caatinga and Pantanal biomes, serves as an ecological corridor for many animal species, making it very rich in biodiversity (BRASIL, 2002).

According to the Ministry of National Integration - MIN (2009), the Cerrado plays an important role in the global ecological balance and is home to approximately one third of Brazil's biodiversity. Since the 1990s, the government and various non-governmental sectors have become concerned with the conservation of this biome.

### 10.1.1 Fauna of the Cerrado

The Cerrado is a biome of great biodiversity, with a great diversity of habitats, which determine a remarkable alternation of species between different phytophysiognomies. 199 species of mammals are known, and the avifauna comprises 837 species. Fish comprise around 1,200 species, approximately 180 species of reptiles and around 150 species of amphibians. There are a large number of endemic fish, but not all of them are known. Amphibians account for around 28% of the biome's endemic species, and reptiles for 17%. According to recent studies, the Cerrado is a refuge for a large number of arthropods, housing around 13% of butterflies, 35% of bees and 23% of termites in the tropics (BRASIL, 2011).

Among the mammals, the following can be highlighted: giant armadillo *(Priodontes maximus),* chicken armadillo *(Dasypus novencinctus),* peafowl armadillo *(Euphractus sexcinctus),* maned wolf *(Chrysocyon brayurus),* giant anteater (*Myrmecophaga tridactyla),* tapir *(Tapirus terrestres),* wild dog (*Tapirus terrestres),* wild dog *(Tapirus terrestres).*
*Cerdocyon thous, Puma yagouaroundi, Ozotoceros bezoarticus, Puma concolor and Panthera onca*. The reptiles are mostly made up of boa constrictors, rattlesnakes, jararacas and teius lizards. Birds include the emu *(Rhea americana),* seriema *(Cariama cristata),* common vulture *(Coragyps atratus),* among others (CAVASSAN, 2006).

The number of herpetofauna species (reptiles and amphibians) may be greatly underestimated, as new species have been described in regions that have not yet been studied by science and there is still a lack of information on amphibian diversity. The richness of reptile species in the Cerrado is quite significant, and is comparable to that of the Amazon when expressed proportionally to the size of the domains, since the Cerrado is only half the size of the Amazon (MACHADO et al, 2004).

The bird class has a significant concentration of species in the Cerrado, with almost half of the species recorded in Brazil occurring in the biome. One of the first major reviews of the Cerrado's avifauna indicated that a large proportion of them (82 %) have some degree of biodiversity.

dependence on forest environments. For this reason, the richness of bird species in the Cerrado is the result of intense exchanges with neighboring forest biomes (Amazon and Atlantic Forest) (MACHADO et. al., 2004).

> With the exception of the majority of birds, the fauna of the Cerrado is generally characterized by its nocturnal and fossorial or subterranean habits, seen as ways of escaping the rigors of the weather during daylight hours. The fauna of the Cerrado is very diverse according to the environment it inhabits. Thus, in addition to the fossorial fauna, there are other groups associated with the litter layer; another with the undergrowth and yet another associated with woody trees. This group includes termites, which build their termite mounds in different shades of color, depending on the type of soil prevailing in the area, making up one of the most typical landscapes of the Cerrado, along with the twisted and suberect stems of many trees (CAVASSAN, 2006, p. 15).

### 10.1.2 The Flora of the Cerrado

In Cerrado areas, due to their territorial extensions and diversity of flora, where there is a consortium of herbaceous, shrub and tree plants, they are home to a fauna rich in diversity and density, whose basis of support is the increase and spatial distribution of the opportunities required by the animals (OLIVEIRA et al, 2002).

The Cerrado is an open vegetation, made up of trees of varying heights, which can reach up to 8 meters, relatively sprawling, whose tops do not touch; sparse shrubs from 0.50cm to 3 meters; herbaceous carpets with a predominance of grasses, mixed with sub-shrubs and some low shrubs. The trees and shrubs are generally crooked, with the cortex of the trunks being quite suberous, cleft, grooved and similar (FIGUEIREDO et al, 2010).

In terms of biodiversity, the Cerrado has diverse ecosystems with a rich endemic flora. It is classified as having primitive vegetative formations, with four divisions: forests, fields, wetlands and humid environments with aquatic plants. The forests occupy the depressions, valleys and watercourses and have few epiphytes. The grasslands cover most of the territory, with formations known as campestres. The vegetation of the marshes is made up of grasses, cyperaceae, shrubs, small isolated trees and some herbs, among other species (MARACAHIPES et al, 2011).

Cerrado vegetation has characteristics that set it apart from adjacent biomes, although many physiognomies share species with other biomes, forming many ecotonal areas. It includes forest, savannah and grassland formations. In forest formations, tree species predominate, forming a continuous or discontinuous canopy. In savannas, trees and shrubs are scattered over a grass layer, without forming a continuous canopy. In grasslands there is a predominance of herbaceous species and some shrubs, with no trees in the landscape (SANO et al, 2008).

## 10.2 AGRICULTURAL EXPANSION IN THE CERRADO

According to Fidelis and Lima (2008), in the last decades of the 20th century, the Cerrado Biome represented 23% of the national vegetation cover, but suffered a loss of approximately 80% as a result of modern occupation by agriculture and livestock. Concern about the remnants of the Cerrado in relation to the introduction of eucalyptus monoculture requires scientific studies to provide answers to society's questions about the drying out and impoverishment of the soil, the decrease in underground water and the loss of biodiversity.

Machado et al (2004) report that occupation of the Cerrado occurred at different times and speeds. Most likely, the opening up of pasture areas for beef cattle was the main cause of deforestation in the Cerrado. Also according to the author, recently the pressure on the Cerrado has started to come from other sources, such as soybean, cotton, corn, millet, sorghum, sunflower and eucalyptus forests.

The suppression of its vegetation for cattle ranching and the cultivation of monocultures such as soybeans, wheat, corn and the planting of forests to serve the pulp and paper industries have completely devastated large areas of the Cerrado. As well as being devastated by plantations, the Cerrado suffers from the fires that occur during droughts (FIDELIS; LIMA, 2008).

When Cerrado vegetation is removed to grow monocultures such as wheat, soybeans or any other type of vegetation that has small, shallow roots, there is a change in the environment. During the rainy season, the roots of these introduced species will not be able to infiltrate the water underground like the roots of the original Cerrado, thus causing the level of the aquifers to drop (BARBOSA, 2013).

According to Nunes et al (2012), soils with original vegetation cover have a better infiltration capacity, due to factors such as the presence of channels formed by roots which are deeper, the presence of organic matter and microbiological activity.

As a result of the advance of the agricultural frontier, deforestation, exotic forests, mining, fire and development projects, around 50% of its original vegetation has been destroyed and disappeared, with the remainder already showing some form of economic occupation with preserved patches (CARDOSO; PIRES, 2008).

> If the history of the great civilizations can be described according to the development of agricultural exploitation and animal domestication, irrigation must be considered as the technology that provided the most prosperity. Irrigation has always been an essential practice for human development, from the natural flood control of the River Nile in Ancient Egypt and the Tigris and Euphrates rivers in Mesopotamia, to the use of animal and human power to raise and distribute water on cultivated terraces in India and China, and the construction of stone canals and embankments by the Andean and Mesoamerican peoples of the pre-Columbian period (FIDELIS; LIMA, 2008).

In the 1950s, the use of new agro-pastoral management was expanded, in this case the extensive use of chemical products to combat pests and insects and increase productivity (MOREIRA et al, 2002). Between 2005 and 2011, the use of agrotoxics doubled in Brazil (RIGOTTO, et. al., 2012), making the country the largest consumer market for agrotoxics; the value of sales reached around US$ 8.5 billion/year (SANTOS, 2012). The abuse of toxic substances (fertilizers, herbicides and pesticides) in agriculture causes a decline in people's quality of life and the destruction of environmental resources (ANDRADE; GANIMI, 2007).

Degradation in the ecological space began with the process of suppressing the native vegetation, which was replaced by the planting of eucalyptus. However, this process was carried out without any study of the negative impacts of the use of streams in the process of suppressing the vegetation, thus not allowing for the selection of species such as pequi *(Caryocar brasiliense),* mangaba *(Hancornia speciosa),* araticum *(Annona coricea*), important species for the livelihood and source of income for small producers (FONSECA, 2008).

Another negative aspect is the contamination of rivers, streams and groundwater by the indiscriminate use of chemical products, and many springs are at risk of disappearing due to the removal of the vegetation that protects them (ANDRADES; GANIMI, 2007).

The adoption of conventional agriculture was a process induced by the state during the military dictatorship, in the 1950s and 1960s, called the "Green Revolution", which contributed to the expansion of soybean crops, increasing income and exports (STOTZ, 2012), and in order to sustain this process, high use of fertilizers and agrotoxics was necessary. Even in the 21st century, when ecological awareness and the search for environmental improvements have grown, it can be seen that the use of agrotoxics doubled in Brazil between 2005 and 2011 (RIGOTTO et al, 2012).

It is worth noting that environmental impact according to CONAMA, Resolution 001/86, e (BRASIL, 1986, p.01)

> [...] any alteration of the physical, chemical and biological properties of the environment, caused by any form of material or energy resulting from the activity which, directly or indirectly, affects: I - the health, safety and well-being of the population; II - social and economic activities III - the biota IV - the aesthetic and sanitary conditions of the environment V - the quality of environmental resources

According to Hespanhol (2002), all human activities, whether industrial or agricultural, must be assessed in accordance with this resolution, so that there can be better environmental quality, providing society with a better quality of life.

> Certain soil management practices, such as intensive cultivation, cause changes in physical and structural properties and influence the erosion process. Monocultures, such as sugar cane, if cultivated for a long time cause physical changes in the soil, especially in its porosity, and intensive eucalyptus monoculture can also cause changes in water resources (BUCKUP, 2006).

## 10.3 ENVIRONMENTAL EDUCATION AND THE PRESERVATION OF THE CERRADO

Observing the impacts and possible depletion of natural resources, government institutions began to look for measures to intervene and mitigate them. In the 1960s, then, there was a greater concern for the environment, bringing the problem of the environment into students' lives, but it was in the 1970s that it gained greater emphasis and came to be characterized as environmental education and included in schools and universities. In Brazil, an important step was taken in relation to environmental education, when it became a requirement in the 1988 federal constitution and was guaranteed by all three branches of government (BRASIL, 2000).

Law No. 9.795 of April 27, 1999, deals with environmental education, establishing a national environmental education policy. The current development model, combined with population growth and the lack of environmental education, has caused ecological disruptions that threaten not only our country, but also our planet (BRASIL, 1999).

EE permeates everyone's participation in an active way, giving them individual and collective responsibility, forming citizens committed to the environmental problems caused by the cerrado biome, due to its biological importance (RUSCHEINSKY, 2002b).

The theme "environment", included in the National Curriculum Parameters (PCN's), is supposed to contribute to the formation of conscious citizens. Citizens who are able to decide and act in socio-environmental reality. Citizens committed to life, to the well-being of each individual and to local and global society (BRASIL, 1998 apud FARIAS; CARVALHO, 2007).

According to Guimaraes (2005), environmental education plays an important role in integrating man with nature. According to him, a harmonious and conscious relationship provides a dynamic balance in nature, through new knowledge, values and attitudes, the participation of students and educators and the whole of society as active citizens in the process of transforming the current reality of cerrado degradation.

The subject is approached in a cross-curricular way in schools, in accordance with the National Curriculum Parameters (PCN's, BRASIL, 2000), seeking to form citizens who are aware of their rights and duties and who are committed to protecting the environment. As an indispensable element in raising people's awareness, environmental education is undoubtedly today the main tool for guaranteeing a future in which everyone lives in total harmony with the environment.

Queiroz et al, (2013) say that environmental education should be introduced into people's lives from childhood in the family and continued at school, because the earlier the process of raising awareness takes place, the greater the chance of forming citizens who are concerned about preserving the environment.

According to Guimaraes (2005), the current situation of environmental gravity in which the world finds itself, as well as Brazil, makes it essential to implement environmental education for the new generations, especially for students in the early years where the process of forming values and attitudes is more relevant, and together with society as a whole, due to the emergency in which we find ourselves.

EE is a proposal that seeks to develop people's knowledge, skills and attitudes towards preserving the cerrado. Nogueira (2013) presents the following concept as a continuous process that seeks to modify values, ways and positions through which society clarifies its opinions on environmental conservation. According to Guimaraes (2005), correct attitudes alone, such as separating materials for recycling, will not be enough to resolve environmental impacts; attitudes and consumerist values need to change.

As a result of the growing environmental problems in the cerrado, which began intensely in the 1970s, the environmental issue began to have a greater focus in terms of ecological control and preservation; stimulating and implementing environmental guidelines in the various development sectors (BRASIL, 2000).

According to Faganha et al (2009), the aim of environmental education is to learn about the relationship between the biotic and abiotic components, among others, that occur in the Cerrado biome, and it is essential to intervene in this area in order to help conserve its biodiversity. This is why it is important and necessary to implement environmental education projects in schools and other environments that strengthen critical and social awareness, so that actions can be taken to help recover and maintain this biome.

Schools and teachers need to have the knowledge and information to be able to work and develop appropriate work with their students. It is believed that the theme of "Environment" should focus on the conservation of the Cerrado. The need for information on the subject of the Cerrado is of the utmost importance for everyone, because this way the process of building and producing knowledge is constant (QUEIROZ et al, 2013).

Environmental issues have emerged in contemporary society, permeating various fields of knowledge and having a significant impact on school education. Thus, environmental education often emerges as one of the main ways of training people capable of dealing with socio-environmental conflicts, people who understand the scientific and technological debates aimed at preserving the cerrado biome (FARIAS; CARVALHO, 2007).

## REFERENCES


ABNT CATALOG. **NBR 12207:1992**. NBR 12.207 1992. Available at: <http://www.abntcatalogo.com.br/norma.aspx?ID=002310>. Accessed on: 06 Oct. 2016.

ANDRADE, D. F. Implementation of Environmental Education in schools: a reflection. In: Federal University of Rio Grande Foundation. **Electronic Journal of the Master's Degree in Environmental Education**, v. 4.Oct/Nov/Dec 2000.

ANDRADES; T.O. GANIMI; R.N. **Revolugao Verde e a Apropriação Capitalist**. 2007. Available at: <> Accessed on: 13 Sep. 2015

AVANZI, D. **Sustainable development and environmental preservation**. 2004. Available at: <http://www.commtacta.com.br/site3/d ownloads/radio/sustainabledevelopmentandenvironmentalpreservation.pdf> Accessed on: 20 Aug. 2016.

AZEVEDO, C. J. C .0 DE. **People's conceptions and practices in relation to household waste in the central area of Uruguaiana, RS**. Uruguaiana, PUCRS- Campus II. Postgraduate monograph. Environmental education. 1996, 68p.

BARBOSA, A. S. "The Cerrado is extinct and this will lead to the end of rivers and water reservoirs" [2013] Goiania GO: **Opgao Newspaper**. Interview with Fernando Leite. Available at < http://www.jornalopcao.com.br>. Accessed on: May 15, 2015.

BEZERRA, A. A. **Fragmentos da historia da educagao ambiental EA**. UFAM. p. 01-06. 2004. Available at: <dialogica.ufam.edu.br /PDF/no3/Aldenice_E ducaga o_ambiental.pdf.> Accessed: July 17, 2015.

BEZERRA, B. G. REIS, L. M. M. **Environmental education at the sewage treatment plant of the federal university of rio grande do norte.** Portal Meio Ambiente UFRN, 2009.

BIERWAGEN, M. Y. **Environmental information in consumer relations**. p. 0117. 2007. Available at: <http://www.rkladvocaia.com/arquivos/artigos/ art_srt_arquivo2 0080731 122009.pdf> Accessed on Aug. 15, 2011.

BRAZIL. **Presentation of the Transversal Themes**: **Environment and Health**. Secretaria de Educagao Fundamental. 2ª ed. Rio de Janeiro. 2000.

. Ministry of Education. Secretariat for Primary Education. **National curriculum parameters**: environment and health. Brasilia, 2000.

. Law No. 9597, of April 27, 1999. Institutes the National Environmental Education Policy.

. National Environmental Council. Resolution n.001, of January 23, 1986.

**Disposes of basic criteria and general guidelines for the Environmental Impact Report (RIMA).** Diario Oficial [da] Republica Federativa do Brasil, Brasilia, DF.1986.

LAW, No. 9.795, of April 27, 1999. **Provides for environmental education, institutes the National Environmental Education Policy and makes other provisions.** Diario Oficial da Uniao, Brasilia, DF, 28 Apr. 1999

. Secretariat for Primary Education. **Parametros curriculares nacionais**: introdução aos parametros curriculares nacionais. Secretariat for Primary Education. Brasilia: MEC/SEF, 1997. Available at:< http://portal.mec.Qov.br/seb/arquivos/pdf/livro01 .pdf>. Accessed on: July 15, 2015

. Ministry of National Integration - MIN 2009. **Strategic Plan for the Development of the Center-West** (2007-2020). Available at: <http://www.integracao.gov.br>. Accessed on: 02/12/2014.

. Ministry of the Environment - MMA 2011. **The Cerrado Biome**. Available at:<http://www.mma.gov.br/biomas/cerrado>Accessed on: 10 Oct. 2015

. Ministry of the Environment - MMA 2002. **Brazilian Biodiversity**. Available at:< http://www.mma.gov.br/estruturas/chm/ files/biodivbr.pdf>Accessed: 10 Oct. 2015.

BUCKUP, L. A. **Monoculture with eucalyptus and sustainability**. 2006. revistas/A_Monocultura_com_Eucaliptos_e_a_Sustentabilidade_.pdf. available at: <http://www.territoriosdacidadania.gov.br/portal/saf/arquivos/view/ater/artigos-e>. Accessed on 01 Aug.2015

BURSZTYN, M. **Ciencia, etica e sustentabilidade**. 2. ed.Sao Paulo : Cortez ; Brasilia, DF :UNESCO, 2001

CARDOSO, R. S. B.; PIRES, L V. **Some considerations on eucalyptus monoculture and its implications.** Federal University of Vigosa - UFV, 2008.

CASCINO, F. **Educagao Ambiental: Principios, Historia, Formacao de professores**. Sao Paulo, Editora Senac, 2003, 3rd Ed.

CAVALCANTI, C. et. al. **Development and nature**: studies for a sustainable society. Institute of Social Research. Recife. p. 262. Oct. 1994. Available at: <http://ww w.ufbae cologica.u fba.br/arquivo s/livro_desv olvimento_nat urez a.pdf>. Accessed on August 20, 2016.

CAVASSAN, O. P. S.; GOMES, P.; SENICIATO, T. **O Ensino de Ciencias, a Biodiversidade e o Cerrado**. Sao Paulo, 2006. Available at: <http://www2.fc.unesp.br/cdmct/arquivos/cursoCerrado/apresentacao.pasta/Cerrado.pdf>. Accessed on: May 30, 2015

CUNHA, S. B.; GUERRA, A. J. **A Questao Ambiental: Diferentes Abordagens.** 4. ed. Rio de Janeiro: Bertrand Brasil, 2008.

DIAS, G. F. **Educagao Ambiental**: principios e praticas. 9. ed. Sao Paulo: Gaia, 2004 .[a]

DIAS, G.F. **Educagao Ambiental: principios e praticas.** Sao Paulo: Gaia, 2004b.

DURAN, N.; ESPOSITO, E. **Effluents and Environmental Microbiology.** Embrapa- CNPMA; Jaguariuna-SP, 1997.

EIGENHEER, E. M. **Solid waste as a subject for environmental education**. ECO*21 magazine. Issue 135. Rio de Janeiro. 2009.

FAQANHA, P. E. W. et al. Teaching about the Cerrado: report of an environmental education experience by the PET/BIOLOGY-UFU group. **In Extension**, v. 8, n. 1, 2009.

FARNESI, C. C.; MELO, C. Educagao Ambiental no Ensino Formal: A Atuagao do Professor. **Educagao** v.27 - n° 01 - 2002.

FARIAS, C. R. O.; CARVALHO, W. L. P. Environmental law in the classroom: meanings of an educational practice in high school. **Ciencia & Educagao** (Bauru), p. 157-174, 2007.

FELIX, R.A.Z. Selective collection in the school environment. **Rev. electronica Mestr. Educ. Ambient**. Botucatu. SP. v.18, jan. to jun. 2007.

FERNANDES, C. **Anaerobic digestion**. 2001. Available at: <http://www.dec.ufcg.edu.br/saneamento/DigeAnae.html>Accessed on: September 25, 2013.

FERNANDES, R. N. A educagao que vem do lixo: Um estudo do programa Piedade sem lixo e a formagao da cidadania ambiental. **Brazilian Journal of Environmental Education.** v.05 n.01, 2009.

FERNANDES, P. A.; PESSOA, V. L. S. The Cerrado and its impacting activities: a look at mining and mechanized agriculture. **Eletrdnica de Geografia**. Uberlandia, V.3, n.7, p.19-37, Oct.
2011.

FIDELIS, A. C.; LIMA, J. D. **Environmental impacts of forestry in the Cerrado biome 2008**. Federal University of Goias - Catalao Campus. Available at:<http://www.uff.br/vsinga/trabalhos/Trabalhos%20Completos/Alex%20Camargo%20Fidelis.pdf>. Accessed on: May 5, 2015.

FIGUEIREDO, M. A. P.et. al. Structural alteration of a cerrado area exploited under a management regime in the municipality of Joao Pinheiro - Minas Gerais-Brazil. **Revista Arvore,** Vigosa-MG, v.34, n.3, p.521-528, 2010.

FLICK. M. E. P. **Environmental Education and Teacher Training**. 2010. Available at:<http://www.cenedcursos.com.br/educacao-ambiental-e-formacao-de-professores.html>. Accessed on 26/08/2016

FONSECA, A. I. A. et al. **The cerrado in the context of eucalyptus monoculture development 2008.** Federal University of Minas Gerais. Accessed on 20/08/2011.

FREIRE, P. **Pedagogy of the oppressed**. Rio de Janeiro: Paz e Terra, 1975.

GADOTTI, M. **Perspectivas actuais da Educagao**. Porto Alegre: Artes Medicas, 2000.

GIORDAN, A. **Basics of Education for Sustainable Development**. In: The Call to Action. Environmental Education Now and for a Sustainable Future (K. Nakayama, ed.), p.137-147, Tsukuba: Tsukuba Shuppankai, 1994. Available at: <http://www.scielosp.org/pdf/csp/v15s2/1294.pdf>. Accessed on March 3, 2011.
GONQALVES, T. V. O. **Ensino de Ciencias e Matematica e Formação de Professores:** marcas da diferença. Campinas, SP p.275, 2000.

GONQALVES, F. S. Interdisciplinarity and the construction of knowledge: a challenging pedagogical concept. **Education & Society**. Campinas, n. 49, p.468-485, dec. 2004.

GRYNSZPAN, D. Health education and environmental education: an integrative experience. Cadernos de Saude Publica (FIOCRUZ), RIO DE JANEIRO, v. 15, n.02, p. 133-138, 1999
GUATTARI, Felix. The three ecologies. Trad. Maria Cristina F. Bittencourt. Sao Paulo: Papirus, 1990.

GUIMARAES, M. **A Dimensao Ambiental na Educagao**. 6th ed. Campinas, SP: Papirus, 2005.

GUIMARAES, R. P.; FONTOURA, Y. R. S. Rio+20 or Rio-20? Chronicle of an announced failure. **Ambiente e Sociedade**. vol. 15. Sao Paulo. 2012 Available at:_<http://www.scielo.br/scielo.php?script=sci_arttext&pid=S1414753X2012000300003> . Accessed on 02/10/2016.

GRYNSZPAN, D. Health education and environmental education: an integrative experience. **Cadernos de Saude Publica**, v. 15, p. S133-S138, 1999.

HIGUCHI, M. I.G. and AZEVEDO, G.C. (2004). **Education as a process of building environmental citizenship**. Brazilian Journal of Environmental Education. Brasilia, DF.: Brazilian Network of Environmental Education. 140p. V. il.; 63- 70.

HESPANHOL, I., 2002. Water reuse potential in Brazil: agriculture, industry, municipalities, aquifer recharge. **Revista Brasileira de Recursos Hidricos**. 7, 75-95.

HUTCHISON, D. **Ecological Education: Ideas about environmental awareness**. Porto Alegre: Artmed, 2000.

IBEMA . **Environmental Education at School**: developing environmental awareness. Available at:<http://www.suapesquisa.com/educacaoesportes/

educacao_ambie ntal.htm>. Accessed on: August 15, 2016.

JACOBI, P, Educagao Ambiental cidadania e sustentabilidade. **Cadernos de Pesquisa.** n. 118, p. 189-205, margo/ 2003. Available at:< http://www.scielo.br/p df/cp/n118/16834.pdf.> Accessed on: Aug. 12, 2016.

JESUS, R. M. et al. Recycling at school: an environmental education action. **Education and Contemporaneity**. Sergipe. Brazil. p. 01-13. Sept. 2011.

KLEIN, A. L. (org). **Eugen Warming and the Brazilian Cerrado**: a century later. Unesp Editora: Sao Paulo - SP, Imprensa Oficial do Estado, 2002.

KLINK, C. A.; MACHADO, R. B. The conservation of the Brazilian cerrado. **Megadiversity,** vol.1, n°1,2005.

KRAEMER, M. E. P. **Social responsibility**: a lever for sustainability. July 2005. Available at: <http://www.gestao ambiental/recebid os/maria_kraemer_pdf/pdf.php> Accessed on Aug. 16, 2016.

LAYRARGUES, P. P. **Socio-environmental conflicts and citizenship: what is the theme of Environmental Education?** In: MATA, S. F. et al. (Orgs.). **Environmental Education: a** commitment to society. Rio de Janeiro: M7 Editora, 1999. p. 50-55.

LIMA, K. E. C.; VASCONCELOS, S. D. **Análise da metodologia de ensino de ciencias nas escolas da rede municipal de Recife** . Rio de Janeiro, v. 14, n. 52, p. 397-412, jul./set. 2006.

LIMA, K. E. C.; VASCONCELOS, S. D. **Essay: evaluation and public policies in education. Analysis of science teaching methodology in Recife municipal schools**. Rio de Janeiro. v. 14, n. 52, jul/set. 2008.

LOPES, A. T. CADEMARTORI, C. V. The study of solid waste recycling in elementary school: a perspective on environmental education. **Revista de ciencias ambientais**. Canoas - Rs. vol. 1, n.1. p. 67-78. 2007.

LOUREIRO, C.F.B. **Trajetoria e Fundamentos da Educagao Ambiental**. 2. ed. Sao Paulo: Cortez, 2006.

MACHADO, R. B. et al. **Estimates of Cerrado area loss**. Brasilia, DF, 2004. Available at: http://www.conservation.org.br/arquivos/RelatDesmatamCerrado.pdf Accessed on: 30.Mai.2015

MARACAHIPES, et. al. Structure and floristic composition of woody vegetation in Cerrado rupestre in the Cerrado-Amazon Forest transition, Mato Grosso, Brazil. **Biota Neotrop,** vol. 11, no. 1, 2011.

MATTOS, N. C. M. et al. Collecting and recognizing plastic: an attitude in environmental education. **Electronic Journal on Environmental Management, Education and Technology REGET-CT/UFSM**. v(1). n°1, p. 01 - 12. 2010.

MAZZINI, A. L. D. **Nosso Lixo de Cada Dia**: desafios e oportunidades. 2. ed. Belo Horizonte: Diretoria de Educagao e Extensao Ambiental, 2012. 64p.
MEIRA, A. M. et al. Environmental education and solid waste: training primary and secondary school teachers. **Revista Cultura e Extensao** - USP. Sao Paulo. vol. 01. p. 01/-88.

MEYER, M. **Educagao Ambiental: pesquisaa-agao.** Ciencias da Educaci6n, Barcelona, v. 6, n. 2, p. 217-231, 1998. Available at: <http://www.scielo.br/pdf/ciedu/v13n2/v13n2a02.pdf>. Accessed on March 3, 2017.

MINAS GERAIS. **COPAM/CERH Joint Normative Decision 01 - Effluent treatment for the state of Minas Gerais, 2008.**

MOREIRA, J. C. et. al. Integrated evaluation of the impact of agrotoxic use on human health in an agricultural community in Nova Friburgo, RJ. **Ciencia e Saude Coletiva**, v. 7, n. 2, p. 299-311,2002.

MOURA, N. Rio+20 leaves more frustrations than legacies. Russian Gazette. 23/06/2012.

NASCIMENTO, A. M. et al. **Waste recycling and green chemistry**. p. 01-71. jul.
- Oct. 2005. Available at: <http://www.ccmn.ufrj.br/curso/trabalhos/pdf/ quimica-trabalhos/quimicameioambiente/quimicaeamb2.pdf>. Accessed on October 3, 2011.

NOGUEIRA, L. M. **Perception of Science Teachers in Education Environmental**: a case study. 2013. available at:< http://www.bioead.com.br/arquivos/monografias/2009 beberibe/Lucindo%20Monteiro%20Nogueira.pdf> accessed on: 10 Aug. 2015

NUNES; J. A. S. et. al. Infiltration Velocity by the Concentric Ring Infiltrometer Method in Cerrado Red Latosol. **Centro Cientifico Conhecer, Goiania, v.8, n.15; p1685 2012.**

OLIVEIRA, A. P. **Turismo e Desenvolvimento: planejamento e organizagao**. Atlas: Sao Paulo, 2002.

OLIVEIRA, A. L. **Educagao Ambiental**: concepções e praticas de professores de ciencias do ensino fundamental. 2007. Available at: <http://www.saum.uvigo.es/reec/volumenes/volumen6/ART1_Vol6_N3.pdf>. Accessed on September 17, 2016.

PANAROTTO, C. The environment and sustainable consumption: some habits that can make a difference. **Revista das Relagoes de Consumo**. p. 132152. 2008. Available at: <http://www.caxias.rs.gov.br/procon/site /_uplo ads/publicacoes /pu blicacao_5.pdf >. Accessed on August 25, 2016.

PAULA, D. **The organization of the Brazilian space**: regions of Brazil. 01/02/2011. Available at: <http://www.slideshare.net/walquiriadu/a-

organization-of-brazilian-space-regions-1-southeast>. Accessed 13 Mar. 2015

PEDRON, N. A., ROSA, M. B. **Environmental education in schools: sensitizing teachers in small municipalities.** Available at: <http://cascavel.ufsm.br/revistas/ojs2.2.2/index.php/remoa/article/view/2778/16 16>. Accessed on March 3, 2016.

PIMENTA, H. C. TORRES, F. M. Sewage: the importance of treatment and technological options. **XXII National Meeting of Production Engineering Curitiba** - PR, October 23-25, 2002

QUEIROZ, H. P. R.; PEREIRA, S. G.; BORGES, D. C. S. A Conscientizacao dos Alunos do Ensino Fundamental para a Preservagao do Cerrado. **Brazilian Journal of Education and Culture.** RBEC. ISSN 2237-3098. n. 8, 2013.

RECH, E. **Biotechnology, sustainability and Brazilian biodiversity**. 2004. Available at: <http://www.cib.org.br/pdf/biotecnologia _sustainability_biodiversity%20.pdf> Accessed on: Aug. 20, 2016.

RIBAS, F. F. Waste recycling: a question of sustainability. **Revista Cientifica Faculdade Modelo** ano I.vol. 01. n. 02. p. 01-28. January-June/2007.

RIGOTTO, R. M. et al. The green economy in the countryside: challenges for research and public policies to promote health in the face of agricultural modernization. **Ciencias de Saude Coletiva**, v. 17, n. 6, p. 1533-1542, 2012. Available at:<http://www.scielo.br/pdf/csc/v17n6/v17n6a17.pdf> accessed on 23 Sept. 2015.

RUSCHEINSKY, A. et al. **Environmental Education**: multiple approaches. Porto Alegre: Artmed. 2002

RUSCHEINSKY, A. **Educagao Ambiental: abordagens multiples**. Porto Alegre: Artmed, 2002 b.

SALES, G.M. **Educagao infantil**: um modo de trabalhar a reciclagem e a preservação do meio ambiente. 2011. Available at:<http://www.artigonal. com/educacao-infantil-artigos/educacao-infantil-um-modo-de-trabalhar-a reciclagem-e-a-preservacao-do-meio-ambiente-4988988.html>. Accessed on: October 4, 2016.

SANTA MARIA, L. C. et al. Selective collection and separation of plastics. **Quimica Nova Escola**. n. 17. p. 32-35. May 2003.

SANMARTI, N. **Educagao ambiental nas escolas**: **reflexoes a partir da experiência na área das Ciencias. Environmental education in schools: new lines of thought and action**. Dossiers. Rosa Sensat 1994.

SANO, S. M. et al. **Cerrado**: Ecology of the Cerrado. Org. Embrapa. Brasilia, DF. 2008.

SANTOS, M. A. et al. **The Brazilian Cerrado**: notes for study. Belo Horizonte,

2010. Available at: <http://web.face.ufmg.br/cedeplar/site/pesquisas/td/TD%20387.pdf>. Accessed on: May 30, 2015

SANTOS, A.S.P. **The importance of sewage treatment** 28/05/2013 . ESA Class 1st Period - The Importance of Sewage Treatment Class 1st Period - ESA, 2013. Available at: < http://www.ufjf.br/engsanitariaeambiental/files/2013 /06/A-Import%C3%A2ncia-do-Tratamento-de-Segos.pdf >Accessed on: 06 Oct. 2 015.

SAUVE, LUCIE. **A cartography of currents in Environmental Education.** In: SATO, Michele e CARVALHO, I.C.M (org) Educagao Ambiental: pesquisa e desafios. Porto Alegre: Artmed, 2005.

SANTOS, L. C. Recycling waste at school. **Forum ambiental da alta paulista**. vol. 4. p. 01-18. 2008.
SANTOS, O. H. dos. SOARES, T. F. et al. A Conversation With Biology Teachers About Teaching Resources. X Jornada de Ensino, Pesquisa e Extensao - Jepex 2010 - UFRPE: Recife, October 18-22.

SCHNEIDER, E. **Municipal environmental management**: environmental preservation and sustainable development. p. 01-09. 1997. Available at: < http://www.portal ga.ea.uf rgs.br/acerv o/ds_art_05.pdf> Accessed on 20 Aug. 2 016.

SHULMAN, L. S. **Towards a pedagogy of cases**. In: SHULMAN, J. H. (Org.) **Case methods in teacher education**. New York and London, University of Colombia, 1992. p. 01-29.

SINO, M. A. **Pet opens the door to new uses.** Available at: **<www**.plastico.com.br/revista/pm342/reciclagem1.htm>. Accessed on April 3, 2017.

SORRENTINO, N. **Environmental education, citizen participation and organization. Em Aberto**, 49:, 1991. p. 46-56.

STOTZ, E. N. Os limites da agricultura convencional e as razoes de sua persistência: estudo do caso de Sumidouro, RJ. **Revista Brasileira de Saude Ocupacional**, v. 37, n. 125, p. 114-126, 2012.

TALAMONI, J. L. B. & SAMPAIO, A. C. (Org.) **Educagao ambiental**: da pratica pedagogica a cidadania. Sao Paulo: Escrituras Editora, 2003.

TEIXEIRA, C. Sustainable Development in Conservation Units: the naturalization of the social. **Brazilian Journal of Social Sciences**. Vol. 20 n°. 59. p.51- 166. Oct. 2005.

TOZONE-REIS, M. F. C. **Educagao Ambiental**: natureza, raza e historia. Campinas, SP: Autores Associados, 2004.

UNICAMP. UNIVERSITY OF CAMPINAS. **The importance of sewage treatment**. Environmental Technologies Teaching Library. Environmental

Sanitation Module. 2013. Available at: <
http://www.fec.unicamp.br/~bdta/es goto/importancia.html>. Accessed on: April 12, 2015.

MIX
Papier aus verantwortungsvollen Quellen
Paper from responsible sources
FSC® C105338

Printed by Books on Demand GmbH, Norderstedt / Germany